The Urbanization Revolution

Planning a New Agenda for Human Settlements

Urban Innovation Abroad

Series Editor: Thomas L. Blair
The Martin Center for Architectural and Urban Studies,
University of Cambridge, Cambridge, England

Urbanization, despite its many severe consequences, has given a healthy stimulus to urban innovation in developing countries. This series seeks to share with an international readership the ideas and experiences of policy-makers, planners, academics, and researchers actively engaged in the day-to-day planning, design, and management of Third World cities.

NEW COMMUNITIES FOR URBAN SQUATTERS: Lessons from the Plan That Failed in Dhaka, Bangladesh
  Charles L. Choguill

STRENGTHENING URBAN MANAGEMENT: International Perspectives and Issues
  Edited by Thomas L. Blair

URBAN INNOVATION ABROAD: Problem Cities in Search of Solutions
  Edited by Thomas L. Blair

THE URBANIZATION REVOLUTION: Planning a New Agenda for Human Settlements
  Edited by Richard May, Jr.

A Continuation Order Plan is available for this series. A continuation order will bring delivery of each new volume immediately upon publication. Volumes are billed only upon actual shipment. For further information please contact the publisher.

The Urbanization Revolution

Planning a New Agenda for Human Settlements

Edited by
Richard May, Jr., AICP
Chair, International Division
American Planning Association

PLENUM PRESS • NEW YORK AND LONDON

Library of Congress Cataloging in Publication Data

The Urbanization revolution: planning a new agenda for human settlements / edited
by Richard May, Jr.
    p.   cm. — (Urban innovation abroad)
  Bibliography: p.
  Includes index.
  ISBN 0-306-43222-6
    1. Urbanization — Developing countries. 2. Urban poor — Housing — Developing
countries. 3. City planning — Developing countries. 4. Urban policy — Developing countries. I. May, Richard, Jr. II. Series: Urban innovation abroad (New York, N.Y.)
HT384.D44U73   1989   89-3861
307.7′6′091724 — dc19   CIP

© 1989 Plenum Press, New York
A Division of Plenum Publishing Corporation
233 Spring Street, New York, N.Y. 10013

Printed in the United States of America

Foreword

Our one world of interdependent peoples is learning that, while our countries are each unique and special, our problems have traits that are general and universal. An outline to study urbanization anywhere will have headings for sections on housing, education, employment, transportation, water supply, nutrition, metropolitan concentration. Everywhere, the large questions will be about sustainable development, rationed resources, human capital and lifting the living standards of people at the bottom.

This book shares information drawn from experience in different countries that are planning for the interlocking complexities of economy, culture and environment. The authors have put many ideas to work and are here reflecting with care on how well they may be working.

The American Planning Association is pleased to have given these papers a forum at our national planning conferences. I thank Richard May for his initiative and energy in calling these contributors together and Thomas Blair for widening their audience with this publication.

Israel Stollman, AICP
Executive Director
American Planning Association

Preface

With the ending of colonial rule following World War II, many new nations came into being with independent governments all sharing the aim of development in the image of their former colonial rulers. Although development has not proceeded as rapidly as hoped by the peoples of the Third World, the problems attendant to development have matured at an increasingly rapid pace. A major source of these problems is the growing rate of urbanization.

Population growth and increased agricultural production achieved by more intensive farming methods have reduced employment opportunities in rural areas. Those losing their livelihood have fled in increasing numbers to the cities in hope of finding employment. With the exception of a few countries like South Korea and Taiwan, industrial growth has not met expectations and has provided employment to only a small fraction of urban migrants.

The urban structure of most developing countries has historically been very limited, consisting in most cases of one primate city, usually the former seat of colonial government and now the capital. Only the largest countries such as India, China or Brazil have a substantial number of large and medium sized cities to absorb portions of the rural exodus. Furthermore, the infrastructure of most cities was designed to serve far fewer people than they have attracted and is less than adequate to also meet the expanding needs of industrial development. Large portions of urban centers have no public water supply and even smaller areas are served by sewers.

Road and public transport systems are severely overloaded and social services for education and health fail to meet requirements where they exist at all.

The first response of Third World governments was an attempt to provide housing for their burgeoning urban population, largely in the form of public projects based on European models established earlier by colonial governments. Governments soon found that such programs were beyond their means and that rental payments affordable by the poor were far below the cost of building and maintaining such projects. Seeking solutions to this dilemma, governments sought assistance from international and national assistance agencies. The United Nations provided the first response in the 1950s by establishing the Centre for Housing, Building and Planning within the Secretariat. The Centre provided technical assistance and training in these three fields, gradually also developing programs in cooperation with the World Bank and national foreign aid agencies which provided capital funding for housing and infrastructure development projects.

Both the funding agencies and the recipient governments soon found that such projects were too expensive in relation to growing needs and were only affordable by the middle class. In fact, most of this housing was taken up by government employees, and was unavailable to those in greatest need. In addition, because of government policies and lax management, rental income was not covering loan repayments, and in many cases the mere cost of housing maintenance.

The vast majority of urban migrants find that central slum areas are already overcrowded and too expensive. Their only resource is to join others invading vacant public and private lands in and near cities, building their own makeshift shelters. Originally, thought of as temporary, these squatter settlements have become permanent communities, housing the majority of urban dwellers in many cities of the Third World. Governments at first viewed these settlements as illegal, unhealthy and a threat to political and economic stability. In the name of slum clearance many of these areas are destroyed by force only to find that the inhabitants moved to other sites or reoccupied them very soon after the bulldozers withdrew.

The lack of significant progress in coping with the urbanization and shelter problem was reviewed in 1976 at the United Nations

Conference on Human Settlements in Vancouver, Canada. The Vancouver Declaration approved at the Conference focussed world-wide attention on these problems, proposed new approaches and called for the establishment of a new organization, the UN Centre for Human Settlements (Habitat) with headquarters in Nairobi and an enlarged staff and budget. In addition, a number of governments and nongovernmental organizations (NGOs) in the industrialized nations agreed to increase their aid in the field of shelter and urban development.

As we enter the fifth decade of international development assistance, coping with urbanization and its impacts on human settlements constitutes the major challenge to planners, administrators and policy makers. There is a demonstrable need to plan not only for the improvement of shelter and living conditions, but to initiate a new action agenda to provide the means by which disadvantaged peoples and regions can be drawn into the social and economic development process. Learning from the successes and failures of past decades and charting new directions for the future is therefore of critical importance.

Faced with a challenge of this magnitude, and in recognition of the United Nations International Year of Shelter for the Homeless, the American Planning Association (APA) gave prominence to human settlements issues at its annual conference in New York, April 1987, and again in 1988 at San Antonio. Seminars, exhibits and plenary sessions organized by the APA International Division provided a forum for an impressive array of American and foreign-based experts to come together to share ideas and outline solutions.

This volume of selected papers illustrates some of the new approaches, many already being tested for broader application, which will form the basis for shelter and urban development assistance during the coming decade. Representatives of the world's leading aid and assistance organizations clearly spell out the new mood of optimism and resource realism which is redefining shelter provision and human settlement improvement activities, and indicate the ways in which this will affect Third World governments seeking to adapt their housing policies to changing economic and institutional conditions. Economists and fiscal analysts describe methods of mobilizing the requisite housing finance, private-public sector cooperation, and urban management skills in the context of sound fiscal policies. Research scholars assist our appreciation of the role of small towns in linking rural and urban development and the need for a more closely integrated physical or spatial planning with the achievement of economic and social objectives. Distinguished experts propose new priorities for foreign consultants, donor agencies, clients and counterparts in technical assistance projects. Since bringing the creative energies of communities more directly into the development process is vital, the volume concludes with case studies of popular self-help and informal sector efforts in shelter and community building, observations on the emerging role of nongovernmental and community-based organizations, and papers on strategies to increase the supply of trained manpower capable of translating shelter and development plans into successful implementation in an integrated and environmentally sound manner. Finally, an eminent environmentalist draws attention to the fragile parameters of global survival and challenges planners to become more aggressively involved in establishing the habitations and productive systems geared to sustain existence within these parameters.

The authors, many of whom hold positions of high executive power and responsibility, are active in international and national aid and assistance agencies, such as the United Nations Center for Human Settlements, the United Nations Environment Program, the World Bank, and the United States Agency for International Development, and are consultants, academics and researchers with extensive field experience in planning for housing and development. From their statements and reports the reader will gain up-to-date information on current policy and practice in the field today; what has or has not worked in the past; what choices and constraints lie ahead, and what changes are deemed necessary in order to achieve feasible and effective solutions.

We owe our deepest thanks to the authors. Their views and interpretations are, of course, their own and should not be attributed to their respective organizations. We have taken the liberty of summarizing and editing many of the presentations and beg the indulgence of our friends and colleagues for any errors that may have been made. Acknowledgements are given to the organizations, institutions, publishers and individuals whose works are quoted or used herein and for use of their illustrations, tables, maps and quotations from their works.

Thanks are also due to the conference seminar organizers and all the participants; their numbers and enthusiasm bear witness to the growing recognition of the importance of the international dimension in urban planning and development studies and practice. Grateful acknowledgement is given to Israel Stollman, Executive Director of the American Planning Association, for his Foreword and support of International Division initiatives to bring the global laboratory of urban experience to the attention of the widest range of planners, both at home and abroad.

The book provides essential reading on housing and urban development issues and we hope it will serve to strengthen the linkages between the many specialists and disciplines who share a common concern for human settlements planning and development. We believe the book will be of serious interest to students, researchers, professionals and academics in universities and technical institutes as an invaluable resource for courses in urban and regional planning, education, public administration, finance and international affairs. Hopefully, the many constructive suggestions offered in the book will merit consideration by national and international agencies, and the non-governmental and community-based organizations, whose policies and actions will shape the planning agenda for human settlements in the coming decades.

Richard May, Jr, AICP
Chair, International Division
American Planning Association
New York, 1988

Contents

NEW ROLES FOR COMMUNITIES, ORGANIZATIONS AND PLANNING EDUCATORS

EPILOGUE: EXPLORING THE PARAMETERS OF GLOBAL SURVIVAL

Shaping the Future: New Commitments to Shelter and National Development Policies

1

Urbanization and Shelter: Policies and Strategies for Developing Countries

Peter M. Kimm
Director, Office of Housing and Urban Programs
US Agency for International Development

A world where even the poorest people are able to live in safe, sturdy housing is a possibility. Whether it becomes a reality depends on a lot of things, but one of them is the application of planning expertise to the shelter development process. My premise is that the solution to the world housing problem does not depend on additional financial resources, but rather on developing and pursuing the right strategy.

What is the housing situation in developing nations? First, rapid urbanization is creating a rapidly rising need for shelter. "Supercities" are emerging. Mexico City is growing at a rate of 1000 people a day. By the year 2000, Sao Paulo, Calcutta and Bombay will each have more than 16 million people. And, early in the next century, urban populations in developing countries will surpass rural populations.

Developing nations will also see a dramatic shift in the nature of poverty. Within 10 years, most of the poor will live in urban areas. Less than 15 years from now, water, sanitation health care and education will be needed for more than one billion additional urban residents. So, urbanization is not an optional matter to be addressed sometime in the future. It is an issue that is shaping the very pattern of national economic growth, the settlement of vast populations, and the social and political stability of many countries in the developing world.

3

The donor community is really only now beginning to acknowl-
edge the profound influence urbanization is having on economic and
social development. A consensus is being formed that improving the
efficiency and productivity of cities and towns is essential to
economic growth. Cities currently contribute over half of the gross
domestic product of developing countries. By the year 2000 they
probably will account for over two-thirds. And as the locus of
poverty shifts from rural to urban areas, efforts to alleviate worldwide poverty must increasingly be directed at meeting the basic
needs of urban populations.

At the US Agency for International Development (AID), the lead
responsibility for addressing these problems rests with the Office of
Housing and Urban Programs. For more than 20 years we have been
providing assistance, principally through our Housing Guaranty Program, which now totals over $2 billion in total loan assistance. We
have experience in more than 50 developing countries, and the
United States is the dominant bilateral donor in the sector.

The Housing Guaranty Program is a mechanism for channelling
private sector loans from the United States into shelter in developing
countries. We provide a US Government guaranty to the private
lender. Most often the direct borrower is a Central Bank, finance
ministry or other governmental entity that provides a host country
guaranty to repay. It gets to use the dollars, and it absorbs the
exchange rate risk. An equivalent amount of local currency is then
made available for shelter financing through intermediary lenders.
Sometimes these intermediaries are national housing institutions, but
we also work with private banks, building societies and savings and
loan associations. AID's Housing Guaranty Program is active all over
the developing world, supporting and working with private lending
institutions in Africa, Asia and the Near East, as well as the Americas.
Our projects are managed by a worldwide staff based in seven
regional offices, working as an integral part of the AID Mission in
any given country. In addition to the $150 million in loans that we
approve each year, we have some grant funds which we use for
training, providing supportive technical cooperation for project
design and implementation, and conducting research and project
development activities.

A quarter of a century ago, developing countries had an image
of appropriate housing based on many things, including what they
saw and learned in the developed countries and what had been built

in many of their cities during the colonial period.   These aspirations led to the construction, sometimes with donor assistance, of housing projects that tended to be built to relatively high standards.   They were built by and through government bureaucracies and sometimes were built to house government employees.   They required substantial subsidies even for the relatively well-off people who would live in them.

The realities of population growth and urbanization made it clear that this approach would not even come close to meeting the housing needs of developing countries.   Even if all donor resources could be devoted to shelter, the problem would still not be solved by this strategy.   So in the 1970s we moved away from project or estate housing and into what we called the "basic needs" strategy.   Our attention turned to providing shelter for the very poor.

To accomplish this, new models were developed: sites and services, slum upgrading and core housing.   These programs allowed the provision of at least minimal services to a vastly larger population.   A sites and services program maps out a housing project, supplying utilities like water, electricity and sanitation facilities. Under this new housing strategy, housing itself initially could be minimal — or nonexistent.   The programs provided families with the opportunity to meet their own basic shelter needs.   Families had their own private space.   For most of them, even the most rudimentary shelter was an improvement over what they had before.

Through this strategy, major accomplishments have been achieved.   The effort is not just to assist in housing a relative few of the poor - it has been to work with developing countries to create shelter programs that are affordable and that hold the potential for meeting shelter needs on a large scale.   AID financed projects emphasize the provision of minimal adequate shelter to the largest possible number of families.   Of equal importance, this assistance is focussed on serving low-income households.   The experience has demonstrated that it is a feasible undertaking — that low-income families can become homeowners and that they are an excellent credit risk.   But even this strategy was incomplete.   Even though more projects were built at less cost, the system still frequently was dependent on government subsidies.   Sometimes land was made available at below market prices.   Sometimes financing was at interest rates that did not reflect the true cost of money.   Or perhaps infrastructure was provided without adequate cost recovery.   And government bureaucracies continued to play too prominent a role in the process. Their direct involvement in lending, construction and other aspects of project design and implementation not only slowed things down, but frequently precluded the market discipline that could assure the most efficient allocation of resources to shelter.

As long as these conditions persisted, new projects would always require new allocations of donor and host country resources, which were simply not available in the quantity needed to address the problem. The clear conclusion was that the key to the solution did not rest exclusively or even primarily with governments — it was to be found in the initiative of the people themselves.

The fact is, the vast majority of the poor in the developing world are being housed. The facilities may be rudimentary or worse, but people are able to obtain some form of basic shelter through their own initiative and through the sometimes mysterious, but clearly effective, workings of what we have come to call the "informal sector". Land was becoming available, credit was coming from somewhere, and housing was getting built through mechanisms which were producing shelter solutions, of a sort, at a far greater rate than government-sponsored programs. The informal sector, as it has come to be known, was probably generating 25 houses for every one generated by a government program.

One of the most important things in the lives of people is where they live. If they move, they will devote a significant amount of time, energy and effort to their new residence. People are prepared to make enormous sacrifices to provide themselves with a decent home. In some low- and moderate-income neighborhoods, people with incomes of $1200 a year own houses worth more than $12,000, more than ten times their annual income.

Putting all these lessons together, a realistic strategy for shelter in the developing world has emerged. The strategy begins with the premise that individual energy is what is needed to solve the problem. The appropriate government response is to solve those problems that the individuals cannot solve themselves. These fall into three basic categories: the availability of land with secure tenure, the provision of infrastructure and the availability of credit. Of these, perhaps tenure is most important. Experience clearly indicates that if individuals have clear title and secure tenure — if they understand that whatever effort they put into their home will be theirs — exceptional amounts of savings are devoted to shelter. This creates security for the family, contributes significantly to capital formation in the country, and increases the health and productivity of its population. Housing finance institutions, cooperatives, private entrepreneurs and above all the individuals themselves will see to it that appropriate shelter is produced, given appropriate government policies.

We have, through our Housing Guaranty Program and related assistance resources, a couple of hundred million dollars a year that we can bring to bear on the shelter and urban problems of the developing world. This is a sizeable sum, but it is a drop in the bucket, given the size of the problem. Nevertheless, we believe our efforts are relevant.

We use these funds to enter into "policy dialogue" with developing countries in an effort to work with them toward strategies that will do the job. Policy dialogue focuses on the national approach to provision of shelter for all. The basic idea is to adopt a series of policies which, in their totality, will set in motion the forces that will put the supply of shelter in some sort of equilibrium with the demand. The goal is to achieve project replicability in the context of a plan to address the larger problem. The most important single consideration is the appropriate division of labors between the public sector and the private sector.

I do not mean to minimize the importance of the government's role. There can be no successful shelter program without appropriate government participation. Our substantial experience in many developing countries argues strongly that government should expend its energy on the things that individuals cannot do for themselves. This facilitative function of government is all-important.

An equally critical element is the encouragement and support of the private sector, which almost everywhere is the principal engine of economic growth, as well as the proven, effective producer of shelter. By providing opportunities to private individuals, as well as developers, contractors and private sector finance institutions, the production of housing for low-income families will be most efficiently achieved.

We work with the countries to adjust their view of what minimum appropriate standards for shelter should be. It is better to meet basic needs by spending less per family so that more families can be served. We work with the countries to develop policies that will assure adequate cost recovery. This is critical to achieve project replicability and produce shelter solutions at a scale adequate to meet the growing needs of the developing world. Full cost recovery means eliminating interest rate subsidies, charging market prices for government-owned land, and recovering through appropriate charges the cost of providing basic infrastructure and services. To the extent that cost recovery is increased, many more families can be served. If government is to absorb some costs, that is, to subsidize, it should do so consciously, considering costs and benefits, politics and equity. Every time AID designs a new Housing Guaranty and then negotiates the loan terms with a host country, these issues are brought to the table.

Experience with subsidies has not been encouraging. Most existing housing subsidies do not reach truly low-income families and do not result in programs of sufficient order of magnitude to really have an impact on the shelter sector. Where subsidies are used, they need to be carefully designed to reach an appropriate target group and to achieve a clearly understood purpose.

On the other hand, there have been individual projects in which attempts have been made to have that group of beneficiaries cover all related direct and indirect costs for services, when this is not yet general practice. This has resulted, in some cases, in the poor paying the full cost and then some, while their more affluent colleagues are subsidized.

What is needed is a balanced approach that results in sufficient income from all sources to permit the critical institutions to function efficiently. Of course, this is easier to say than to do. Finally, we try to use our assistance to help beneficiary countries improve the institutions, public and private, that effect housing programs and urban development.

One aspect of improved shelter programming that deserves special note is the need for greater attention to the role of women. They often carry an exceptional burden in the cities of developing countries. Many provide primary economic support for their families frequently as entrepreneurs. At the same time, they are responsible for the crucial and time-consuming tasks of child-rearing and household management. They are also often important community leaders who are instrumental in working with urban agencies that provide services to their neighborhoods. It is critical, therefore, that the design and implementation of shelter improvement projects account for women's needs and take advantage of their special knowledge and skills. One way to accomplish this is for women to be able to take increased professional responsibility for project development. This is a goal which AID has been pursuing for many years.

We strongly encourage the building and strengthening of institutions which serve people locally and in which they can participate. We are supporting both neighborhood and cooperative organizations, and we are working with governments to encourage decentralization. We believe, very deeply, and have figures to support this belief, that in a single generation, perhaps 25 years — within the limits of the resources that are now available — an adequate, if bare-bones house for every family on the face of the earth is a reasonable, achievable goal. This can be achieved if the right policies are followed.

2
Human Settlements in National Development Policy: The Year 2000 Agenda

Darshan Johal for Dr Arcot Ramachandran

Executive Director, United Nations Center for Human Settlements (Habitat)

High population growth rate and rapid urbanization are the twin forces accentuating the crisis of human settlements in many developing countries. Soon the population of our planet is expected to surpass the five billion mark. Most of these five billion people are already living in the developing countries and in substandard settlement conditions from which there are not easy prospects of escape. At the same time, the process of urbanization is currently transforming settlement patterns in developing countries at an unprecedented pace. In 1950 less than 300 million people lived in towns and cities in developing countries: by 1985 this number had swollen to 1.1 billion and by the year 2000 there will be approximately two billion urban inhabitants in the developing countries, with a sizeable proportion concentrated in a number of agglomerations of tens of millions of people, the so-called mega-cities.

What stands out in the area of institutional trends is that most developing countries have not yet been able to devise adequate institutional systems to face this crisis. Their building standards, regulations and codes are unrealistic in relation to the prevailing economic, social and climatic conditions and prevent the poor from improving their shelter and neighborhoods. Their human and material resources need to be developed and mobilized on a scale corresponding to the magnitude of the crisis.

To meet this challenge, the United Nations General Assembly designated 1987 as the International Year of Shelter for the Homeless

(IYSH). This decision represented a growing international recognition of the magnitude of the global shelter crisis and the need to address it urgently. It was also recognized that the practical action would have to be taken primarily at national and local levels.

The program for the Year can be divided into three phases. The preparatory phase began in 1983 and ended at the close of 1986. It was a period in which countries established national agencies to co-ordinate their program for the Year, reviewed their national shelter policies and programs and identified the areas most in need of action. To date, 140 countries have identified several hundred demonstra-tion projects as a means of finding new solutions to shelter issues and a great many countries are trying to find more effective ways of meeting shelter needs. The International Year itself (1987) repre-sented a period of review, information exchange and policy develop-ment. The third phase stretches to the year 2000.

Over 400 IYSH projects have been designated, each of them highlighting the crucial issues in the provision of shelter and services for the poor. The new policies and programs of public agencies include the granting of tenure in several countries in Latin America and many upgrading projects — especially in the cities of South and Southeast Asia. Private initiative plays a role in numerous upgrading and sites-and-services projects reported by almost all countries. Urban management issues are being tackled in projects to strengthen institutions in a number of countries, and several countries report increased attention to the financial and institutional arrangements, affordability and cost recovery. Popular participation is a common thread running through most of the projects whether public or private, nationwide or specific, urban or rural.

The United Nations Center for Human Settlements (Habitat) is the United Nations agency responsible for coordinating action in the field of human settlements; it is also the secretariat responsible for coordinating the IYSH program. The Center is currently carrying out 167 technical cooperation projects in 85 developing countries, and has conducted extensive research work to find practical solutions to shelter problems, assist governments to develop affordable policies that meet local needs and to use legislation as a tool for positive change. Our views on the shelter crisis in developing countries have, therefore, not been developed in isolation sitting behind a desk, but are the result of many years of field work in the slums and squatter settlements of Africa, Asia and Latin America.

At the tenth session of The United Nations Commission on Human Settlements in Nairobi in April 1987, the Executive Director, Dr Arcot Ramachandran, presented a New Agenda for human settlements and a special program to provide shelter and services for the poor. Global in scope, but with focus of action at the national and local level, the agenda attracted overwhelming support from 106 member states and other intergovernmental and nongovernmental organizations represented at the session.

A major feature of the New Agenda is to counteract the view of human settlements as a social expenditure which detracts resources from productive uses. Investments in the development of human settlements must be seen as an essential prerequisite that provides a framework for social and economic development. Therefore, national human settlements policies should become a central concern of the national development process, rather than the preoccupation of isolated human settlement agencies and institutions. Only through such integration can human settlements development derive the dimension and the effectiveness called for by their role in national development.

Given the continuing trends towards urbanization, and particularly, metropolitanization, in the developing countries, new processes of human settlement of development will have to be put in place. Traditional concepts of long-range master planning, land-use zoning, development control and so on will have to be abandoned and replaced by a view of human settlements management as a coordinated process for planning, building, operating, maintaining and renewing the settlement fabric.

An important component of the new commitment to action must be the improvement of the shelter conditions of the urban and rural poor. Governments generally committed to "national development" must be persuaded that no real development can be achieved when large percentages of the population have to live in inhuman conditions. They must show their commitment by setting specific targets to redress this situation. Explicit commitment to human settlements in national development policy should be made a prerequisite for international cooperation.

However, given the limited resources available and the scale of the problems, governments committed to the improvement of the living conditions of the poor and disadvantaged can only "melt the top of the iceberg". While recognizing the uniqueness of individual national situations, governmental and intergovernmental institutions should consider the merits of replacing previous approaches, limited to negative control, and fostering the mobilization of efforts of all potential contributions, private and public alike.

Such strategies should apply to the whole spectrum of the human settlements development process. In terms of population distribution policies, emphasis should be directed at exploiting the expected rather than planning the unfeasible: instead of formulating policies aimed at limiting urban growth, efforts would be best directed at diffusing the urbanization process and its beneficial effects in such a way as to create real alternatives to metropolitan growth through the promotion, for example, of networks of intermediate settlement systems capable of exploiting the potential of urbanization in various regions.

On the detailed level, enabling strategies imply conceiving and formulating styles of management capable of exploiting the potential for public-private partnership. While the public sector would profit from adopting the principles of ingenuity, effectiveness and accountability, the formal private sector should have an opportunity to show responsibility for an involvement in human settlements development and investment. Finally, enabling strategies should recognize and support the informal private sector. This calls for elimination of barriers now stifling efficient market functioning and a strong government program to encourage modernization and improvements in the operation of the informal private sector.

At the institutional level, enabling strategies would imply priority attention to the potential of decentralization and to the appropriate balance of functions with the role of central government concentrated on coordination guidance and monitoring and that of subnational and local government planning, execution and operation. This is not to say that governments should withdraw from their responsibilities related to management of human settlements development, mobilization of financial resources and the needs of the poor. On the contrary, by focussing efforts on these prime responsibilities, instead of spreading resources too thinly over peripheral activities, governments will be able to give urgent problems the attention they deserve. If governments are to take maximum advantage of all existing capacities and potentials, they need to consider the following specific steps:

1.   Formulate a policy framework, emphasizing human settlements contribution to expanding production, creating employment opportunities, and generating economic growth.

2.   Take an integrated approach to human settlements development, involving community participation including, especially, the participation of women as a cardinal principle.

3.   Develop human resources both in the public and private sector.

4.   Generate financial resources, providing incentives and opportunities for the formal and informal private sectors, and mobilizing domestic household savings.

5.   Develop new partnerships for the delivery of shelter and infrastructure.

6.   Promote the construction sector by giving primary attention to the development of indigenous building materials and to the elimination of institutional barriers preventing the local building materials industry from responding effectively to the demand.

The global strategy for shelter to the Year 2000 was strongly supported by the Habitat Commission and resulted in a draft resolution submitted to the General Assembly. Under this resolution, the General Assembly decided that there shall be a Global Strategy for shelter to the Year 2000, including a plan of action for its implementation, monitoring and evaluation. All governments are urged to commit themselves to the objectives of the Global Strategy by adopting and implementing shelter strategies designed to enable the mobilization of all forces and resources for the attainment of the objectives of the strategy. They are urged to renew this commitment annually by announcing on World Habitat Day the concrete actions to be taken and targets to be achieved during each successive year. All United Nations bodies and agencies and the international community at large are requested to support the formulation and implementation of the Global Strategy.

Briefly, the points emphasized in the Guidelines for shelter strategies include:

1.   Shelter strategies should be an integral part of development strategies.

2.   Shelter strategies should take into account the multidimensional nature of the problem and fulfill and reflect the wide socio-economic benefits of shelter development.

3.   Criteria of affordability and replicability, particularly for shelter for low-income population group, should be reflected and special attention should be paid to improving the access of the poor to land, building materials and finance.

4.   All efforts should be made to involve, in full partnership, all relevant governmental, nongovernmental, public and private sector organizations, in particular, the communities and people concerned, in the planning and implementation of shelter strategies.

5.   Special attention should be paid to the problems faced and capacities represented by marginal groups such as women, and youth, and disadvantaged groups such as the aged and the disabled.

6.   Governments should report biennially at the session of the Commission on Human Settlements on the progress made in implementing these measures.

This is the task for all of us concerned with issues of human settlements between now and the Year 2000.

Mobilizing Resources for Urban Housing, Infrastructure and Finance

3
Policy Options for Urban Housing in Developing Countries

Dennis A. Rondinelli
Senior Policy Analyst, Research Triangle Institute

Increasing the access of the poor to adequate shelter is one of the biggest challenges facing governments in developing countries as cities continue to grow and larger numbers of poor households continue to concentrate in them over the next two decades. In order to meet the growing needs of the poor for shelter, governments will have to explore a wide range of options for housing construction, financing and a land acquisition. New combinations of policies will be needed that include strengthening the private sector's capacity to provide affordable housing, and supporting self-help housing construction. This paper examines the policy options through which governments can assist in providing adequate shelter for the growing number of poor households expected to be living in cities in the future.

POLICY OPTIONS FOR NATIONAL SHELTER STRATEGIES

A variety of options must be pursued by governments in developing countries to provide adequate housing to urban residents, and among them are: slum clearance and public housing, sites-and-services and slum upgrading schemes, government assisted self-help housing construction and improvement, private and informal sector housing construction, cooperative housing schemes, modification of land use and building regulations, employment and income generating programs for the poor to increase their ability to pay for housing, and land tenure programs. (See Table 1). Each of these options has advantages and limitations, and the experience with each option must be carefully assessed in planning national shelter strategies.

Slum Clearance and Public Housing

Governments in developing countries have in the past sought to remove the poor from slum neighborhoods and rehouse them in low-cost shelter. In most countries, however, slum clearance programs have accomplished neither goal. Indeed, in some countries such policies have exacerbated the problems of the poor (Von Einsiedel and Molina, 1985; Seguchi, 1985; Hashim, 1985; Whang, 1985). The failure of these policies led many governments in developing countries to try massive public housing construction during the 1960s and 1970s. The problems of the poor were defined primarily by the condition of their housing, and the solution was to construct public units with relatively low rentals. But in most of these projects neither services nor employment opportunities were usually provided, and the results were equally disappointing. The cost of public housing construction was high and rentals were expensive. As a result, these policies usually benefitted middle-income rather than the poorest families. Most slum dwellers were merely pushed from cleared sites to other parts of the city (Rondinelli and Cheema, 1987).

The failure of slum clearance and public housing policies to deal effectively with the problems of slum dwellers or to provide other services needed by growing numbers of poor urban households became clear by the early 1970s (World Bank, 1980; Kulaba, 1982). Among the most serious defects of these policies are that they are extremely costly for national governments because of the high level of compensation paid to owners of demolished properties and the high cost of construction; they create serious problems of social dis-placement and disruption for the residents of slum and squatter settlements; and they are often delayed by social and political pressures exerted by slum residents, who resist forced removal from their homes. They often impose high transportation costs on families who are relocated far from their workplaces in the center of the city, and thus rarely alleviate the housing problems of most of the poor and, indeed, exacerbate them in many countries. The poor cannot afford much of the public housing that replaces slum dwellings and thus the destruction of slum communities often reduces the stock of low-income housing and worsens overcrowding in low-rent units.

Table 1. Policy Options for Increasing Access of Poor to Shelter

Delivery Options	Policy Options	Financing Options	Land Acquisition Options
Public			
Direct provision by national, municipal or local governments	Slum clearance and public housing	General revenues Special revenues Leveraging of assets through borrowing	Eminent domain and compensation Advance acquisition and land banking Government purchase Bartering and exchange Land use controls
Provision by public corporation, enterprise or special authority	Land use and building regulations		
Private			
Provision by private firms, individuals or cooperatives	Cooperative housing Self-help construction Private construction	Contributions and donations Indirect subsidies Cross subsidies Private financing Mobilization of savings	Private purchase Land readjustment
Informal sector provision	Contract construction		
Mixed			
Government assisted self-help	Sites-and-services and core unit upgrading	Cofinancing Mobilization of government revenues to increase access of poor to credit	Land use controls Land readjustment Bartering and exchange Gifts and donations
Coproduction	Subsidized private housing construction		
Public sector financing or subsidy of private sector provision			

Although public housing for the poor still plays an important role in most national shelter strategies, experience suggests that it alone is too costly and limited in scope to meet the shelter needs of poor households. Public housing policies must be combined with other options if they are to have a serious impact on reducing shelter deficiencies.

Sites-and-Services Programs and Provision of Core Housing

Beginning in the 1970s, many governments in developing countries sought alternatives to public housing for meeting the shelter needs of the poor. One extensively used alternative is the provision of low-cost core housing units, which poor families can upgrade or expand as their income rise. A closely related option is the sites-and-services program, in which government housing agencies assemble, clear, and provide with basic infrastructure, land that is divided into home sites. Poor families build their own basic shelter, usually with subsidized materials or with credit provided at low rates of interest, or contract with small construction enterprises to build basic dwelling units for them.

Sites-and-services policies of the 1970s and 1980s were designed to make shelter and community services affordable for the poor by introducing them incrementally, at standards that kept costs low, or by having community groups contribute labor, money and materials. Many countries developed sites-and-services, upgrading, and integrated shelter and urban services projects with assistance from the World Bank and other international funding agencies.

Some observers estimate that sites-and-services projects can provide appropriate housing for a cost of about three to five times less than public housing. In addition, they are often more beneficial because they allow poor households to keep monthly payments low and to improve their housing only when they have accumulated sufficient resources. Sites-and-services projects allow the poor to pursue their own housing priorities, to contribute to the construction of their dwellings, and to use locally available building materials (Merrill, 1977). Sites-and-services and upgrading programs can be effective means of providing access for the poor to shelter, especially when they are combined with self-help construction policies, but they must be carefully planned and efficiently administered if they are to achieve their full potential. Among the most important forms of support are the provision of basic infrastructure and utilities, land tenure, and low-cost credit.

Evaluations of World Bank-assisted sites-and-services and slum upgrading projects in Asia, Africa and Latin America indicate that after basic services are provided in the area, nearly all poor families eligible for building material loans took out the full amount to which they were entitled. They built more quickly when they were allowed to arrange for construction of their own dwellings, and many used small-scale contractors and hired help in addition to contributing their own labor. Moreover, in nearly all projects, the improvements in housing were substantial and the amount of savings mobilized by dwellers was much higher than analyses of their monthly incomes indicated would be possible (Mwono, 1978; Strassman, 1982; Keare and Parris, 1982).

The World Bank's experience with sites-and-services and up-grading projects has been that although these approaches are more effective in providing affordable shelter for the urban poor than slum clearance and public housing programs, most sites-and-services schemes remain pilot projects serving a relatively small percentage of the poor households needing shelter (World Bank, 1983). Five major obstacles to expanding sites and services appropriate to shelter provision have been identified (Cohen, 1983): (1) inadequate numbers of trained professionals who can design, facilitate, and manage effectively sites-and-services projects; (2) difficulties in developing and providing inexpensive building materials and technology that the poor can afford; (3) reluctance on the part of public bureaucracies to elicit community participation on the design and implementation of the projects; (4) ineffective information dissemination among developing countries about those methods and techniques of self-help construction that work best; and (5) weak financial institutions for providing low interest loans needed by the poor to improve and expand their dwelling units.

Keeping projects affordable and still recovering costs has been especially difficult during periods of high inflation. Cost recovery has been impeded by rising costs of materials and equipment, forcing the government to lower the standards of services provided to poor communities or requiring slum dwellers to pay a larger percentage of their household incomes for service improvements than do higher income groups (Von Einsiedel and Molina, 1985).

*Government Assisted Self-Help Housing Construction
and Improvement*

To meet the rapidly expanding shelter needs of the poor in developing countries, governments must also explore policies that promote self-help by the poor on a much larger scale than has been possible through sites-and-services projects.   Either in conjunction with sites-and-services and upgrading schemes, or through individual efforts, self-help housing construction is the primary means by which the poor obtain shelter.   In some countries, government assistance to self-help efforts have been successful in allowing large numbers of the poor to build core dwelling units (Laquian, 1983).

Experience in Colombia indicates that the poor will build their own houses or will hire informal sector builders for low wages to assist with the more difficult aspects of construction, if government provides land tenure and basic public services on the housing site.   It was found in Colombia that "the best way to assist low cost self-construction of housing is to provide building materials at a reasonable price" (Sanders, 1983: 7).   Exploitation of the poor by merchants and the expense of transportation drove up the cost of building materials and became a serious obstacle to self-help efforts in *barrios* in Cali, for example, until the Carvajal Foundation provided building material warehouses in poor neighborhoods.   The warehouses sold construction materials to *barrio* residents at market prices rather than at the inflated prices of private merchants.   Competition from the warehouses forced local merchants also to sell materials at market prices, thereby increasing the amount of materials available to *barrio* residents.

Experience in many other developing countries indicates that self-help approaches have a number of advantages (USAID, 1963). Among the advantages are that it: (1) makes shelter available for those low-income families who have no other means of securing decent housing; (2) reduces the cash outlays of poor households for housing construction by as much as half because labor costs often account for a large portion of total construction costs; (3) promotes greater pride and satisfaction of home ownership because of direct participation of poor families in house construction; (4) increases the real wealth of poor families without encouraging or causing inflation; (5) helps develop building skills among low-income people in countries where construction skills are needed; (6) may increase the demand for building materials that can be produced by small-scale local industries; and (7) encourages personal interest in home maintenance and expansion after construction is completed.

Heavy reliance on assisted self-help as a deliberate government policy to provide shelter for the poor requires some degree of organization and public promotion, however, and can also have some disadvantages as a means of meeting shelter needs quickly. Among the possible disadvantages are that it: (1) requires the commitment of participants' time and labor, for which there may be many competing demands; (2) requires strong, and often organized, effort to maintain the initial enthusiasm of participants throughout the home building process; (3) requires the development of building skills by people who may have no further need for them except for maintaining their homes, after their dwellings are completed; (4) often leads to slower rates of home construction than by the contract method; and (5) can often result in a lower quality of building construction than if the dwelling were built by more skilled workers.

Despite potential drawbacks, self-help housing construction is likely to remain the primary means by which the poor are sheltered. Government policies that assist or facilitate self-help can make it possible for poor families to obtain at least basic shelter. But experience suggests that to work effectively, self-help efforts must be supported by government agencies, which have often been reluctant to work closely with community groups. In many countries, making self-help assistance programs more effective requires changing the attitudes of political leaders and public administrators about their roles in service provision, and creating new incentives and career rewards for professionals and technicians to respond to the needs of the urban poor. In some countries, it may also require restructuring the responsibilities of community service delivery organizations, and strengthening linkages between public agencies and private organizations, the informal service sector, and community groups (Rondinelli and Cheema, 1987).

Cooperative Housing Construction

Policies promoting cooperative housing construction are another means of supplementing public and self-help shelter programs. In communities where residents cannot easily obtain credit, shelter can often be provided through mutual benefit organizations, in which residents pool their resources to buy materials and contribute labor to construct each member's dwelling. Members of the cooperative usually assist each other in constructing core units until all members have core shelter, and then help each other to expand or upgrade their houses as they acquire the resources to do so.

Assessing the experience with cooperative housing policies, Guhr (1984) found the following advantages of this approach. First, cooperative housing programs can help to create integrated urban communities, not only for the purpose of providing housing, but also for supplying services and facilities and promoting employment opportunities and education, thereby raising the community's standard of living. Second, cooperatives provide internal control to prevent speculation and illegal sale of houses. Third, cooperatives create collective systems of financing and repayment and reduce the dangers of default by instilling principles of mutual responsibility in their members. Fourth, cooperatives allow members to assume gradually the responsibilities for managing housing construction activities and administering the organization, thereby reducing costs. Fifth, cooperatives help mobilize savings among, and create resources for self-help activities by, their members. Sixth, cooperatives provide an efficient arrangement for collective maintenance of houses and neighborhoods. Finally, the savings to poor households from constructing a house through a cooperative can be substantial. For example, cooperatives in Maseru, Lesotho that received financial assistance from the United Nations to establish a revolving mortgage fund were able to reduce the building costs of their members by about one-third (Altmann and Baldeaux, 1981).

Experience with cooperative shelter programs indicates that financial and technical assistance are needed from government or private sources. The members must be willing to work together, and they must have a minimum amount of capital and income to repay their loans. The group's resources must be carefully managed so that all members can share in the benefits.

Governments can support housing cooperatives in a number of ways. First, they can assist in organizing cooperative organizations and in making their advantages known to potential members. Second, governments can provide training in cooperative organization and operation and well as in techniques of housing construction. Third, governments can make extension agents available to work with cooperatives during the construction and maintenance phases. Fourth, they can help mobilize "soft-loan" capital to initiate the revolving loan fund that can be replenished when housing construction is completed and the loans are repaid. Fifth, governments can assemble land on which low-cost shelter can be built by the cooperatives and transfer title to homeowners. Sixth, they can play an important role in assisting cooperatives to obtain reasonably priced building materials. Finally, governments can support cooperative housing by extending basic infrastructure and services to the housing sites.

Reviewing the experience with housing cooperatives in Africa, Altmann and Baldeaux (1981:51) conclude that they "are valuable tools for self-help development provided that they are used appropriately, that is, not overburdened by unrealistic expectations, nor unacceptable for ideological reasons, nor entered into as easy-going solutions". The cooperative housing option should be seen as one of a number of alternatives for meeting the shelter needs of the urban poor.

Private and Informal Sector Construction

In most developing countries, the construction industry provides only a small fraction of the total number of housing units needed, and the poorest households can afford very little of the housing constructed by private firms (Rourk and Roscoe, 1984). Not only does the private sector fail to meet the housing needs of poor households, but it usually fails to serve middle-income families as well (USAID, 1983a). The informal sector rather than the home construction industry provides much of the housing in developing countries, and especially for the poorest households (Cooperative Housing Foundation, 1984; Hardoy and Satterthwaite, 1981).

Increasing the access of the urban poor to decent and affordable shelter depends to a large degree on the ability of the construction and building materials industries to meet their growing needs. Governments can assist private and informal sector construction enterprises to keep building costs low in poorer neighborhoods by opening building supply outlets that sell construction materials at market prices, and to obtain access to land by acquiring and servicing sites that can be subdivided into small parcels that small-scale developers can afford to acquire. Government can also help small-scale construction firms to participate in public housing construction, sites-and-services projects, and upgrading schemes by phasing the projects into incremental, smaller scale, sub-projects (Liebson, 1982).

In many developing countries the capacity of the building materials industries to provide low-cost construction materials, and of the indigenous construction industry to deliver affordable housing, must be expanded substantially. The United Nations Center for Human Settlements (1984) suggests that governments give increased attention to: (1) formulating and enforcing appropriate standards for the production and use of building materials; (2) altering construction and building regulations to allow the use of low-cost materials that provide acceptable levels of performance; (3) supporting the testing of new and locally available building materials that lower housing costs; (4) the expansion of indigenous capabilities to perform construction work by developing appropriate craft skills and assisting small- and medium-sized specialty and contracting firms; and (5) developing programs to assure that small- and medium-scale construction firms have access to working capital and opportunities to bid competitively on public housing and shelter projects.

Policies have been enacted in some countries to require private companies locating in large cities to provide staff housing or rent-subsidies for workers, and this is another means by which governments can induce the private sector to produce more housing. In Bangkok, Thailand for instance, some companies provide land at factory sites on which workers construct houses for themselves and their families. Other companies provide dormitories for young single workers, and some railroad companies use their large tracts of land to build barracks-type housing in which workers and their families can live (Angel, Benjamin and DeGoede, 1977). Regulations requiring large companies to provide housing or subsidies for employees to secure housing is one means of shifting the social costs of industrialization to the industries that create new demands for shelter, but they can also make the cities that adopt them more costly locations for private firms.

Modifications in Land Use and Building Regulations

In addition to improving the capacity and efficiency with which national, metropolitan and local governments provide housing for the urban poor, and encouraging private industry and self-help housing construction, governments in developing countries can also lower the costs of shelter by changing land use and building regulations to make them more appropriate to developing country needs. Ironically, municipal governments in many developing countries adopt building codes and land use standards from Western industrial countries that are not only inappropriate to local conditions, but that create unnecessary construction problems and raise costs. Overly restrictive housing construction standards place home ownership beyond the means of low- and middle-income families. By lowering density controls, lot coverage, room floor area requirements, and height controls, for example, in ways that do not endanger human health or safety, municipalities can increase housing production, make units less expensive, and make land use more efficient, thereby lowering the costs of extending public services.

Controlling land uses, land prices, and speculative practices can also reduce the costs and increase the access of the poor to community services. Rapidly rising land prices and land speculation practices drive up the costs of housing construction and price poor families out the market. The explosive costs of land has not only made it difficult for lower income families to obtain decent housing in central cities, where many have jobs or are engaged in informal sector activities, but drives many low-income residents from the core of cities to their peripheries. The separation of living and work areas not only increases the strains on public transportation services, and raises the commuting costs of the poor, but also increases the costs of extending services to peripheral neighborhoods where the poor live. Rapidly rising land values increase costs for small enterprises in cities, push people from the center to the fringes of urban settlements, accelerate the conversion of agricultural land to urban uses, promote sprawl, and increase the costs of acquiring rights of way for public utilities and property for schools and other public facilities.

One means of lowering land costs for housing is through public acquisition and sale of land to cooperatives, private companies, government housing authorities, and individuals who will construct shelter that is affordable for the poor. In most developing countries, land acquisition costs are among the most expensive components of housing construction. Among the policy options available to governments in developing countries for acquiring the land needed for shelter and community facilities are (Kitay, 1985): (1) purchase and reservation of property through land banking to assure its availability at affordable prices for low-cost housing construction or infrastructure and service provision; (2) public acquisition of leasehold interests and options to buy land that may be needed for low-cost housing; (3) adoption and enforcement of appropriate land use regulations and controls to assure adequate land for housing and infrastructure and services in areas of the city in which the poor live;

(4) land readjustment programs that take a portion of property from private developers to recover the costs of service provision or to use for low-cost housing, infrastructure, or public facility construction; (5) land bartering and exchange either with private owners or among government agencies for appropriate property for low-cost housing; (6) land confiscation through eminent domain to acquire sites for public housing, facilities, and infrastructure; and (7) gifts, contributions and donations of land by private owners for low-cost housing construction or public services, in consideration of tax advantages.

Employment-Generating Programs That Increase Income and Housing Demand

Perhaps one of the most important ways that governments can increase the access of the poor to shelter is to promote employment generation programs that raise the incomes of the poor sufficiently to create greater effective demand for housing. Although there is often an inadequate supply of housing for the urban poor, the solution may not be entirely on the supply side, but in raising the low levels of effective demand among the poor due to their low incomes.

This policy can be implemented by designing public housing and service improvement programs to generate as much employment as possible for their beneficiaries. UNICEF's basic services strategy, for example, attempts to improve urban services in ways that will build the skills and raise the incomes of people — and especially of women — living in the neighborhoods where the services will be delivered. Neighborhood women have been trained, for example, to help run day-care and preschool centers, and community residents have been trained as paraprofessional health workers to staff neighborhood clinics and family planning centers. (UNICEF, 1982).

Governments can also develop programs that increase the capacity of the informal sector to provide appropriate services, build low-cost housing, or provide construction materials. The informal sector is an important source of income for the urban poor in most developing countries, and with proper support could construct housing and provide more community services in low-income neighborhoods. Some forms of housing improvement, small-scale transport, water supply, and public safety could be provided through the informal sector at a lower cost than by municipal governments (Montgomery, 1984). Employment can be generated for the poor by designing public housing and facilities projects to use indigenous materials and components, such as pipes, electrical accessories, cement blocks, bricks and lumber, that can easily be produced by small-scale enterprises in the area where housing will be constructed, and that use local contractors and labor.

Increasing the Security of Land Occupation

Finally, increasing the security and stability of land occupation for the urban poor is one of the most important ways in which governments can promote self-help housing construction and service improvement programs.   In the past, governments have taken action against squatters and slum dwellers through legislation, eviction, and relocation.   They have developed upgrading schemes that do not transfer title to land.   These actions invariably increase the insecurity of squatters and slum dwellers and undermine the effectiveness of house upgrading and community development programs.   Experience suggests that when squatters to not have the security of land tenure, it is extremely difficult to initiate or sustain self-help activities. Without secure rights in land occupation, slum dwellers are not moti- vated to contribute their time, money, and energy to upgrading their dwellings.

The best means of creating security of land occupation will, of course, differ from country to country.   Two successful approaches have been land readjustment and sale of land to squatters through national mortgage banks.   Provision of dual systems of tenure — community ownership of land and family ownership of dwellings — combined with sites-and-services programs, is also an effective way of securing land occupation in some communities (Rondinelli and Cheema, 1986).

IMPROVING THE IMPLEMENTATION OF SHELTER POLICIES

Over the next decade, governments will have to assess more carefully the best combination of policy options to meet the housing needs of the poor.   Inevitably, national housing strategies will have to include both public housing construction and programs that encourage poor families to improve their housing incrementally through self-help activities.   But designing more effective national housing strategies will do little good unless governments also improve the implemen- tation of housing policies.

Much more attention needs to be given to identifying the differing needs and characteristics of the urban poor and to tailoring housing programs to them. Poor families living in the same neighborhoods rarely have homogeneous characteristics. Recent rural migrants often have different problems obtaining shelter than longtime residents, and the employed poor can afford different kinds of housing than those who cannot work. Moreover, housing policies often have drastically different impacts on the various economic, social, ethnic, religious, and cultural groups. Construction of standard housing units often ends up being inappropriate for some and ineffective in meeting the needs of others. More precise and accurate identification of the needs of the poor can contribute to more effective, efficient, and relevant housing policies (Sinclair, 1978). The most appropriate mode of providing housing for the urban poor depends on a variety of factors such as the size of the city, the number and characteristics of the urban poor to be served, the cohesiveness of neighborhoods, the characteristics of the shelter needed, and the degree of national and local political commitment to meeting housing needs (UNICEF, 1982; Rondinelli and Cheema, 1987).

Finally, governments will have to find new ways of raising the financial resources needed to meet even the basic shelter needs of their growing urban populations. In most developing countries governments have to do more to strengthen the capacity of private financial institutions to provide mortgages to low- and middle-income families.

Acknowledgement

This paper is a revised version of material included in the author's report *Shelter, Infrastructure and Services for the Poor in Developing Countries* for the United Nations Center for Human Settlements, and the conclusions are those of the author and should not be attributed to any organization.

REFERENCES

Altmann, Jorn and Dieter Baldeaux, 1981, Cooperative housing in Lesotho, South Africa, *Ekistics*, No. 286:49-52.
Angel, Shlomo, Stan Bejamin and Koos DeGoede, 1977, The low income housing system in Bangkok, *Ekistics*, Vol. 44, No. 261:79-84.

Cohen, Michael A., 1983, The challenge of replicability: Towards a new paradigm for urban shelter in developing countries, *Regional Development Dialogue*, Vol. 4, No. 1:90-99.

Cooperative Housing Foundation, 1984, "Contribution of the Informal Housing Sector to the Construction of Housing," US Agency for International Development, Washington, DC.

Guhr, I., 1984, Cooperatives in State Housing Programs - An alternative for low income groups? *in*: "People, Poverty and Shelter: Problems of Self Help Housing in the Third World," R.J. Skinner and M.J. Rodell, eds., Methuen, 80-105, London.

Hardoy, Jorge E. and David Satterthwaite, 1981, "Shelter: Need and Response," John Wiley and Sons, Chichester.

Hashim, Shafaruddin, 1985, Policy issues in urban services for the urban poor in a plural society: The case of Malaysia, United Nations Centre for Regional Development, Nagoya: Japan.

Keare, Douglas H., and Scott Parris 1982, Evaluation of shelter programs for the urban poor: Principal findings. *World Bank Staff Working Papers*, No. 547, World Bank, Washington, DC.

Kitay, Michael G., 1985, "Land Acquisition in Developing Countries: Policies and Procedures of the Public Sector," Oelgeschlager, Gunn and Hain, Boston, MA.

Kulaba, S. M., 1982, Housing and self-reliance in Africa: The role of manpower, regional and in-country training institutions, *in*: "US Agency for International Development, Eighth Conference on Housing in Africa," USAID:61-68, Washington, DC.

Laquian, Aprodicio, A., 1983, "Basic Housing: Policies for Urban Sites, Services and Shelter in Developing Countries," International Development Research Center, Ottawa.

Liebson, David 1982, Development and entrepreneurial capacity, *in*: "Eighth Conference on Housing in Africa," USIAD:115-16, Washington, DC.

Merrill, Robert, N., 1977, Projects and objectives for sites and services, *in*: "Low Income Housing-Technology and Policy," Vol. III, Asian Institute of Technology:1167-1172, Bangkok.

Montgomery, John D., 1984, "Improving Administrative Capacity to Serve the Urban Poor," United Nations Centre for Regional Development, Nagoya, Japan.

Mwono, Nathaniel T. K., 1978, Qualifying and selecting the participants in a sites and services project, *in*: "Fifth Conference on Housing in Africa," US Agency for International Development, USAID:81-85, Washington, DC.

Rondinelli, Dennis A. and G. Shabbir Cheema, 1985, Urban Service Policies in Metropolitan Areas: Meeting the needs of the urban poor in Asia, *Regional Development Dialogue*, Vol. 6. No. 2: 170-190.

Rondinelli, Dennis A. and G. Shabbir Gheema, eds., 1987, "Urban Services in Developing Countries: Public and Private Roles in Urban Development," Macmillan, London.

Rourke, Phillip W., and Andrew D. Roscoe, 1984, "An Assessment of National Housing Needs and Affordability in Kenya," 1983-2003, USAID: Washington, DC.

Sanders, Thomas G., 1984, Promoting social development: private sector initiatives in Cali, Colombia, *USFI Reports*, No. 21, Hanover, NH: University Field Staff International.

Seguchi, Tetsuo, 1985, "Policy Issues in Urban Services for the Slum Dwellers in Madras, India," United Nations Centre for Regional Development, Nagoya, Japan.

Sinclair, Stuart W. 1978, Housing preferences and the urban poor in less developed countries, *Ekistics*, No. 270:243-45.

Strassman, W. Paul, 1982, "The Transformation of Housing: The Experience of Upgrading in Cartagena," Johns Hopkins University Press, Baltimore, MD.

UNICEF, 1982, "UNICEF Urban Basic Services: Reaching Children and Women of the Urban Poor," E/ICEF/L.1440 Add. 1, UNICEF, New York.

United Nations Centre for Human Settlements, 1984, "The Construction Industry in Developing Countries," Report HS/32/84/E, UNCHS/HABITAT, Nairobi, Kenya.

US Agency for International Development, 1963, "Leader Training for Aided Self-Help Housing," USAID, Washington, DC.

US Agency for International Development, 1981, "Morocco Shelter Sector Assessment," USAID, Washington, DC.

US Agency for International Development, 1983, "Urbanization and Urban Growth as Development Indicators in AID-Assisted Countries," USAID, Washington, DC.

US Agency for International Development, 1983a, "Urbanization Trends and Housing Conditions in Central America," USAID, Washington, DC.

Von Einsiedel, Nathaniel and Michael L. Molina, 1985, "Policy Issues in Urban Services to the Poor: The Case of Metro Manila," United Nations Centre for Regional Development, Nagoya, Japan.

Whang, In-Joung, 1985, "Policy Issues in Managing Urban Services for the Poor: The Case of Squatter Improvement in Seoul Korea," United Nations Centre for Regional Development, Nagoya, Japan.

World Bank, (1980), "Shelter: Poverty and Basic Needs Series," World Bank, Washington, DC.

World Bank, 1983, "Learning by Doing: World Bank Lending for Urban
    Development 1972-82," World Bank, Washington, DC.
Yeung, Yue-Man, 1985, Provision of urban services in Asia: The Role
    of People-based mechanisms, *Regional Development Dialogue*,
    Vol. 6, No.2: 148-163.

4
Developing National Housing Strategies: Lessons Learned from Barbados, Jamaica, Jordan and Kenya

Raymond J. Struyk
Principal Research Associate, The Urban Institute

The magnitude of the housing problems in developing countries is well-known in general and in the past few years they have been quite accurately documented in about twenty nations using the Housing Needs Assessment Model developed by the US AID Office of Housing and Urban Programs. A recent study estimates that developing countries as a group must produce about 45 million additional units of minimally acceptable quality each year in the years immediately ahead if they are to meet their housing needs. The rough estimate of the corresponding annual investment is $130 billion or about 5.8 per cent of their combined Gross National Product. Low-income countries, as defined by the World Bank, must produce two-thirds of the housing units at a cost of about $24 billion (Struyk, 1987, Annex E).

To address these staggering problems countries must develop effective, realistic strategies for the housing sector. Failure to do so may well mean that scarce resources are wasted on expensive "false starts" by initiating inappropriately expensive or misdirected housing programs. In the past, few developing nations produced such strategies, but recently several countries have undertaken comprehensive housing strategy development efforts, generally with significant donor assistance.

This paper briefly describes the experiences of four countries: Barbados, Jamaica, Jordan and Kenya, and distils some lessons from them for ways of organizing and conducting effective and imple-

mentable housing programs.   The essential aspects of the strategy
development process are highlighted.

THE FOUR COUNTRIES

The four cases represent an extremely interesting mix along two
germane but different dimensions: the basic attributes of the
countries which affect their current and future housing circumstances
and the processes employed in developing the strategies.   In terms of
basic attributes, there are wide differences among the four countries
in several key indicators: levels of economic development, urbanization, and housing needs.

Selected Indicators	High	Low
Annual population growth	4 per cent (Kenya)	1.2 per cent (Jamaica)
GNP per capita (1984)	$1,570 (Jordan)	$310 (Kenya)
Urbanization	65 per cent (Barbados)	18 per cent (Kenya)
Housing quality	Good (Jordan)	Poor (Kenya)

Kenya, for example, has both a very high population growth rate
and poor quality initial housing, implying the need to produce a high
volume of minimally adequate housing for new families and to
upgrade much of the existing stock.   By contrast, Barbados has a relatively good initial housing stock and low population growth rate.
Obviously, the strategies appropriate for these diverse circumstances
are quite different.

Turning to the divergence in the broad process used to develop
the strategies, Table 1 gives a few simple indicators.   Shown are relative rankings of the size of the effort (in terms of total person
months), the extent of expatriate consultant involvement, the degree
to which government participation in the effort was concentrated in a
single office or spread over a number of agencies, the length of the
process from initiation of work until completion of a full draft strategy, and the current status of the strategy.   As can be seen from the
entries, there are widely different combinations of these factors
represented, however, each of these exercises involved the same
series of steps ranging from the documentation of the existing housing conditions to the vetting of an initial strategy.

Table 1. Selected Indicators of Process of Developing Four National Housing Strategies

Country	Size of Effort	Extent of Expatriate Involvement	Participation By	Time for Development[1] (months)	Current Status[2]
Barbados	Large	Extensive; long-term	Single agency	24	Being implemented
Jamaica	Moderate	Small; short-term	Widely shared	8	Adopted
Jordan	Large	Extensive; long-term	Single agency	18	Under review
Kenya	Moderate	Moderate; short-term	Widely shared	6	Under review

1. Months between formal start of process and preparation of full draft policy statement ready for dissemination beyond the agency managing the process.
2. As of June 1987.

KEY ELEMENTS

Each of the strategy development efforts contained the common elements described below.   While these are presented as discrete,
sequential segments for expositional clarity, the reality was typically
a much more integrated process.

Government Involvement

Each of the four countries decided that the development effort would
be worth it.   However, the countries differ dramatically in the degree
to which they were responding to donor pressures, the amount of
attention from senior officials at the project's start, and the clarity
and comprehensiveness of the charge given to those undertaking the
actual strategy development.   There are some fairly stark contrasts
here among the four countries — for example, between Jamaica,
where intense high level government attention was clearly present
and Kenya where it was lacking.

Obviously, variation in these factors will effect the success of the
overall effort.   Most desirable is a commitment at the ministerial
level and a clear statement of the strategy's objectives expressed at
the outset.   Examples of the desired objectives — beyond increasing
the volume of acceptable and affordable housing — are a better
targeting of public resources on the lowest income households, an
increase in the role of the private sector generally in housing production, or making greater use of the informal sector in securing housing
production.

Establishment of a broad-based steering committee to oversee
the work in principle and to garner support for the effort proved useful in a couple of cases.   The steering committee was composed of
high ranking persons; day-to-day direction fell to a task force; and
technical work was assigned to a series of working groups composed
of experts drawn from the public and private sectors.   Clearly critical
is creation of a strong, well-staffed government body for managing
and assisting day-to-day development activities.

Needs Assessment and Contrast with Current Production

The initial analytic step is to determine the country's housing needs —
both currently and for the next 15-20 years.   For this purpose, the
Housing Needs Assessment Model or the similar model developed by

the Finnish Ministry of Environment were used. (Struyk, 1987a; Lujanen, 1986; Habitat, 1984). These models provide estimates of the number of new and upgraded dwellings needed over the planning period; they also indicate the investment program necessary to produce these units, given explicit assumptions about minimum dwelling standards. Simulations of investment requirements (and implied government subsidies to meet needs) under different housing standards always led to considerable discussion and ultimately a decision about this difficult issue.

An additional aspect of the output is that total investment is divided among land, infrastructure, and the dwelling proper. Requirements for housing finance can also be readily derived from the figures produced by the models.

The estimates just described can be contrasted with estimates of current production: for example, the volume of serviced plots developed annually; the number of units developed, both total and those meeting minimum standards; and the volume of formal housing finance. It is very desirable to examine differences between needed and actual production by household income group, if possible. These differences provide a good measure of the task that must be accomplished and serve as targets for the rest of the strategy development process. This task is efficiently done by a small team of analysts who can share their results with others working on the strategy in the form of a background paper (for a good example see Jones and Turner, 1987).

Sector-by-Sector Assessments

This activity involves the analysis of bottlenecks and constraints to the necessary levels of production in each sector and the development of recommendations for alternative ways of overcoming these. The relevant sectors include land, infrastructure, finance, government housing programs, the residential construction industry and building materials production. Here it is essential to disaggregate among the problems faced in production for different income groups (market segments) and to develop corresponding solutions. In terms of building up the production of minimally acceptable units for lower income households, it makes sense to concentrate on increasing the production from individual households and small entrepreneurs who constitute the informal sector.

This activity seems to be well executed by a series of task forces — one for each sector — which include private and public representatives working together.   Staff resources, often in the form of local consultants, are essential.   Each task force should produce a report stating the constraints and ways to deal with them.

Putting It All Together

Now is the point to move toward an overall strategy.   This is the most technically difficult step, since it requires going beyond the mere integration of the individual pieces developed by the task forces to define an overarching strategy.   It requires a basic understanding of the local scene and how housing markets work and a genuine vision of alternatives to current institutional arrangements.   The product is a document for widespread discussion consisting of: (a) a background statement of housing needs, production deficits, constraints; (b) a statement of objectives, general principles, and the broad strategy; and (c) recommendations for changes from the status quo on a sector-by-sector basis.

In Kenya, for example, the objective was a rapid and sustained increased production of low-cost housing.   One of the principles of the strategy was to assign the private sector the primary production role, with government discharging more the role of facilitator than direct producer of housing.   This really meant a greater role for the informal sector, especially the small investor, in rental housing.   The package of recommendations, therefore, included: (a) a sharp reduction in subdivision standards to realistic levels; (b) the servicing of private land by local authorities for use by homeowners and small investors, since individual investors would not be able to undertake subdivision-level investment on their own; and (c) the encouragement of formal housing finance institutions to make medium-term loans to the small investors.   This brief outline suggests a hallmark of a solid strategy: that it be internally consistent and the individual recommendations reinforce each other.

The task of developing the strategy is well handled by a small drafting committee of three to six persons.   At least two of the members should have a good overview of the entire housing market. Expatriates with wider experience can be especially useful in the forum.

Getting The Decisions

The objective at this stage is to obtain decisions among the options developed previously, while at the same time keeping the central thrust of the strategy together. A two-phase process is almost certainly necessary. In the first, there is wide consultation among those interested, both from government and the private sector, and a genuine attempt should be made to get small scale developers and builders to participate. Government participants should include those from the ministries of finance and planning and the central bank as well as from the housing industry itself to build a broad base of support for the strategy.

This phase can be handled well in a one- or two-day seminar or workshop at which the proposed strategy is vetted and the specific recommendations discussed. It is important that the seminar be carefully planned and that the minister or senior official of the agency responsible for the development of the strategy chair the meeting and be actively involved in inviting the attendance of other senior persons both private and public.

The second phase consists of a round of consultations at high government levels in which final decisions are taken, based in part on the discussion at the workshop. For example, in Kenya the Ministry of Housing, Works, and Physical Planning held discussions with the Ministry of Settlements and Land and the Ministry of Local Governments about their roles under the new strategy. It is wise to share the final draft of principles and recommendations with the office of the president or prime minister to be certain that there is agreement from this quarter. Once this agreement is reached, the process proceeds to its final step. Good examples of the kind of documents describing the strategy at this stage are Government of Jamaica (1987) and Government of Kenya (1987).

Refinement and Initial Implementation

Now is the time to refine the strategy so it can be promulgated as a White Paper, Sessional Paper, or the equivalent document for formal cabinet approval. It is at this stage that annual numeric goals are added and more detailed assignment of responsibilities for particular aspects of the strategy execution are made. Once the paper has been approved by cabinet, then the hard, detailed planning and legislative work necessary to begin implementation is begun. Government of

Barbados (1986) is an example of a document containing detailed numeric goals.

SOME LESSONS LEARNED

Participating in several housing strategy development efforts has revealed some clear lessons that may be useful to those initiating such exercises.

Follow the Market

Given the volume of housing needed and the stringency of the resource constraints facing developing nations, the appropriate role for government is to facilitate the production of housing by the private sector. Local and national governments should concentrate on those tasks which individuals and small entrepreneurs find hard or impossible to do. Acquiring blocks of land and servicing it for private, small-scale development probably heads this list. In carrying out such facilitating actions, it is absolutely essential for governments to move with the market. Perhaps this dictum is best explained with an example.

In Kenya 80 per cent of the housing in urban areas is rental. There are very strong urban/rural linkages which make many of the migrants to the cities think of themselves as only temporary urban residents, although they may remain in the cities for a decade or more. Moreover, there is a tradition of semi-communal living arrangements to which the rental market — especially small investors — has readily responded. In contrast, the donor community has consistently funded the development of site-and-service schemes for owner-occupants. Evaluations of several projects have shown that the typical pattern has been for these projects to become predominantly rental after only a few years. The failure of these assisted projects to "follow the market" has led to a good deal of wasted resources through an inefficient transfer process in which allottees make a large capital gain from investors who purchase their units. The possibilities for improving this approach are numerous and are recognized in the country's draft housing strategy statement. (See Government of Kenya, 1987; for more details on the operation of the rental markets in Kenya, see Struyk and Nankman, 1987).

It's a National Strategy

There is a natural tendency for housing strategies to focus their attention on urban areas. Typically much more data are available for urban areas and the donor community will have been active there, producing reports, analyses and suggestions for innovation and reform. Likewise, Government often views the problems of rural areas to be so immense as to be beyond its ability to offer significant improvement. Nevertheless, the widespread housing problems of rural areas, and the fact that Government is often already extensively intervening through housing-related activities such as water supply projects, suggested that at a minimum the broad outlines of a strategy should be formulated.

To draw on the Kenya case again, the government sought to focus its activities in rural areas on the provision of infrastructure and for the coordination of government-sponsored infrastructure projects now being carried out by a number of different agencies. It also called for establishment of a housing improvement loan program, which would make small loans at market rates. Government funds would be used since private banks are currently unwilling to make medium-term loans in rural areas. The loans would, however, originate and be serviced by private bank branches to introduce them to this type of lending; and, as private institutions began making loans from their own funds in a particular region, government funds would be shifted to other underserved regions. Eventually, the government program would be discontinued.

It Must be a Local Creation

If the strategy is to be truly accepted and implemented, it must be perceived as the country's own creation. While members of the donor community and consultants can help — and should argue strongly for their concepts — ultimately everyone wants the decisions to be made in a thoughtful fashion by senior government officials who have been thoroughly engaged in the process. The further the strategy deviates from local ideas and the lower the involvement of senior officials, the more likely the strategy will be a statement which is never implemented. From the outset, Government must be enthusiastically behind such an exercise.

In summary, development of a successful housing strategy is a difficult and demanding task. Nevertheless, the potential payoff in terms of the volume of housing production meeting minimum standards — especially for lower income households — and the increased efficiency of market operations can more than reward the effort.

Acknowledgement

Some of the work reported in this paper was supported by the United Nations Center for Human Settlements and the USAID Office of Housing and Urban Programs; however, the views expressed are strictly the author's own.

REFERENCES

Government of Barbados, 1986, "Barbados: The National Housing Plan, 1985-1989," Housing Planning Unit, Ministry of Housing and Lands, draft, Barbados.
Government of Jamaica, 1987, "Jamaica: National Shelter Sector Strategy Report," author, Statement presented at the International Year of Shelter for the Homeless meetings, April 1987, Kingston.
Government of Kenya, 1987, "National Housing Strategy for Kenya, 1987-2000," Department of Housing, Ministry of Works, Housing and Physical Planning, Statement Presented at the International Year of Shelter for the Homeless Meetings, April 1987 Nairobi.
Habitat, 1984, "Guidelines for the Preparation of Shelter Programs," United Nations Centre for Human Settlements and the Ministry of Environment, Finland.
Jones, E. B., and Turner, Margery, 1986, "Jamaica Shelter Strategy: Phase I Final Report," The Urban Institute, Report to the Government of Jamaica and USIAD Office of Housing and Urban Programs, report 3666, Washington, DC.
Lujanen, M., 1986, "Spreadsheet Program for the Preparation of Shelter Programmes in Developing Countries," Ministry of the Environment, draft, Helsinki.
Struyk, R., 1987, "Assessing Housing Needs and Policy Alternatives in Developing Countries," USAID, Office of Housing and Urban Programs, US Contribution to the International year of Shelter for the Homeless, Washington, DC.
Struyk, R., 1987a, Planner's notebook: The housing needs assessment model, *Journal of the American Planning Association,* Vol. 53, No. 2:227-34.
Struyk, R., and Nankman, P., 1986, "Developing a Housing Strategy for Kenya: Recent Housing Production, Market Development and

Future Housing Needs," The Urban Institute, Report to Government of Kenya, UN Centre for Human Settlements, and USAID Office of Housing and Urban Programs, paper 3660-1, Washington, DC.

5

Major Trends in Housing Finance and Implications for Developing Countries

Mark Boleat

International Union of Building Societies and Savings Associations

There have been major changes in the housing finance markets of most industrialized countries in recent years. Generally, there has been a closer integration between housing finance and other financial markets, the role of specialist housing finance institutions has been declining, the housing finance function itself has been fragmenting, and there has been a move towards greater use of variable interest rates. As in so many other areas, there is a general presumption that developing countries have much to learn from the experience of industrialized countries, but all too often there is a failure to appreciate the very significant differences between the two groups of countries. This paper surveys the major developments in housing finance techniques in the industrialized countries and then considers what relevance they have, if any, in developing countries.

MAJOR TRENDS IN HOUSING FINANCE IN INDUSTRIALIZED COUNTRIES

In industrialized countries two basic types of housing finance system can be identified:

1. The deposit-taking system, whereby the funds for house purchase loans are raised predominantly from the retail savings markets and the institutions which collect those savings make the loans. One variant of this system is the contract scheme by which entitlement to a loan follows a period of contractual savings.

2. The mortgage bank system, whereby house purchase loans are made by institutions which raise their funds on the capital markets.

Among the institutions which use the deposit-taking system are commercial banks, savings banks, and specialist institutions such as savings associations in the USA and building societies in the UK and Australia. The distinguishing characteristics of these three groups of institutions are:

1. Commercial banks conduct a full range of retail, commercial and, generally, international banking business and are likely to have not more than 20 per cent of their assets in loans for house purchase.

2. Savings banks deal predominantly with the retail sector, although increasingly with business activity as well, and may have up to 50 per cent of their assets in loans for house purchase.

3. Building societies, savings associations and similar institutions concentrate their lending on house purchase and housing related purposes and have 60-80 percent of their assets in house purchase loans.

A few years ago it was possible to draw fairly rigid lines between particular types of housing finance system and housing finance institutions, but recently these lines have become increasingly blurred and there has been a convergence between housing finance instruments and institutions. The reasons for this convergence are part of a worldwide trend towards despecialization of financial institutions and a closer integration of all the financial markets. This has not happened by any grand design, but rather through natural economic forces. Technology has tended to break down barriers between financial institutions. Technology has, for example, made it easier for institutions to offer a complete retail banking service without any branches through the use, for example, of automated teller machines. Technology has made it possible for intermediaries such as real estate agents to originate mortgage loans directly on to the books of a remote lending institution. Technology has also facilitated movements of large sums of capital between institutions and between countries. Generally, technology has reduced the importance of branch networks as a means of distributing financial products such as mortgage loans.

Technology has also contributed to a breaking down of the regulatory barriers which have become increasingly ineffective. The

USA is a case in point. The housing finance institutions were inhibited from competing across the full range of financial services by legal constraints on their activities. Accordingly, the Congress passed the Garn/St Germain Act which freed savings associations and in effect removed the distinction between them and savings banks.

There are parallel developments in other countries. In the UK, for example, from the beginning of 1987 building societies have been acting under a new Act of Parliament. The new Building Societies Act allows building societies for the first time to lend without security and they can also offer financial services including money trans-mission, credit cards, cheque books, foreign exchange services, personal pensions and personal equity plans, and insurance interme-diation. Mortgage banks have had to face a similar competitive threat. In countries where savings banks and retail institutions generally have been prohibited from making long-term loans these constraints are gradually being removed, partly because technology has made them unworkable.

As well as these forces, there has been a trend for financial institutions to diversify their sources of funds. Traditionally, savings associations and building societies have raised their funds entirely from retail sources. However, new techniques in the financial markets have made wholesale funds attractive, in particular the Eurobond market and floating rate instruments generally. Mean-while, the cost of retail funds has been bid up and administration costs have remained high. In the USA there has been the development of a huge secondary market, albeit primarily because the primary market is so manifestly inefficient. In other countries traditional retail institutions have also resorted to the wholesale market. In Britain, until 1983, building societies had no significant power to raise non-retail funds. In that year they were permitted to raise certificates of deposit for the first time and from 1985 they could raise money on the Eurobond market. Subsequently, over a third of the new money has been raised from the wholesale markets and building societies have found themselves able to take advantage of money for five or seven years carrying a variable interest rate a small fraction over money market rates at a time when their retail money has been costing them considerably more.

The internationalization of the financial markets generally has made it possible for American housing finance institutions to raise funds on the international markets backed either by mortgages or, more commonly, by government securities or government guarantees of mortgage loans.

Some deprecate these trends, seeing them as the end of housing finance and implicitly assume that housing finance can be catered for adequately only by specialist institutions dedicated to this particular market. In a sophisticated economy this line of reasoning is wrong. Of all financial products in a well-developed economy a housing finance loan is one of the simplest, and it does not take any great expertise to put a loan together or to sell it and the customer has virtually no brand loyalty. He is far more concerned about where his money is deposited rather than from where he borrows.

The despecialization of financial institutions is thoroughly desirable and in the interests of the consumer. In Britain, for example, despecialization has led to house buyers having a choice of institutions from which they can obtain loans including the traditional building societies, the commercial banks, and a range of new institutions including American banks. Perhaps paradoxically, despecialization has been accompanied by a fragmentation of the housing finance function, with the financing part being separated from the origination and servicing part. The small all-purpose institutions have tended to lose out to the large conglomerates and small specialist institutions.

A second common trend in housing finance throughout the world is the greater use being made of the variable rate mortgage. Those countries which were dedicated to fixed rate mortgages have finally realized what problems this instrument causes. The customer should not be expected to guess what mortgage rates will be doing and therefore when he should take out his loan. There is also strong evidence that the fixed rate mortgage accentuates the natural cycles in the housing market. If interest rates are perceived to be high, then people hold off buying and developers hold off building. If interest rates are perceived to be low then there is a demand which could well lead to over-heating in the market. This stop-go cycle is apparent in the American mortgage market which has been dependent on fixed rate loans. Interestingly, some of the recent research in the USA suggests that the use of the adjustable mortgage rate over the past few years has moderated the cyclical swings in housing market activity.

IMPLICATIONS FOR DEVELOPING COUNTRIES

All of this has little relevance to developing countries. Whether loans are made at fixed or variable rates of interest, whether they are made by a specialist or general institution, and whether there is a secondary mortgage market, is largely irrelevant in a country like India where there simply is not enough money to make loans, and even when loans can be made there is no adequate system of mortgage security and even if that exists, most people cannot afford to buy a house anyway. The problems of developing countries are very different and need to be analyzed quite separately from the problems of industrialized countries.

It is all too easy to look at the size of the problem and say that developing countries have such massive housing problems that nothing can be done about them. This would be to adopt a very defeatist attitude. There are many imaginative housing and housing finance projects in developing countries, often assisted by the international aid agencies such as the World Bank, the International Finance Corporation, and the United States Agency for International Development. Generally, there is a consensus as to what needs to be done in respect to housing finance in developing countries.

One agreed point is that the government generally should not be involved except to create the right framework. In most countries there is not the right framework for housing finance systems to develop. In India, for example, it is impossible to gain possession of a property from a borrower who has defaulted within ten years. The intention of this law is, of course, to protect home buyers. The effect is precisely the opposite. If a lending institution cannot take possession within ten years then it will be very wary about the people to whom it lends and in India the normal practice is not only to take the property as security but also to require two guarantors for any loan. This naturally makes the house purchase process a lengthy one and some people who could afford a mortgage loan cannot obtain the necessary guarantors and therefore cannot borrow. Also, if eventually the lending institution does have to take possession of a property it is bound to make a significant financial loss and this will affect its overall costs and therefore the rate at which it can lend money.

In many other developing countries a critical problem is security of land tenure. A loan secured by a mortgage on a property carries a much lower rate of interest than a loan which does not have benefit of such security. If people are living in property or building their own house or improving their own house and they do not have a legal title to that property, then it will cost them far more to borrow money than if they do have a legal title. What one finds is inefficient land title systems and prohibitions on subdividing plots, all of which again are designed to protect the consumer but which have the opposite effect by not recognizing the limits of what people can afford.

Generally, in developing countries there is also a lack of a sophisticated financial system. Most people do not put their money in financial institutions because they are wary of them and often with good cause. The major financial institutions in many developing countries are government-run savings banks which all to often have handed over all of their funds to the government. Rates of interest paid by financial institutions have been held down to unrealistically low levels. A low mortgage rate or low lending rate generally is perceived to be good for the borrower and there is therefore the implicit assumption that investors should be subsidizing borrowers. Holding interest rates at an artificially low level inevitably has led to a shortage of funds. This is something which regulators and policy makers in a large number of countries have failed to realize.

Government can create the right climate for housing finance institutions to thrive by providing an appropriate land title system and by allowing interest rates to reflect market forces. However, even with these developments it would be unrealistic to expect a massive growth of housing finance institutions willing to provide funds to people wishing to build or improve their own homes. Some pump priming is often necessary, and significantly the aid agencies have concentrated very much on institution-building rather than on providing direct loans for house purchase. There are sadly few examples of really effective housing finance institutions in developing countries. Perhaps the best one is the Housing Development Finance Corporation in India, an institution which was founded with the help of money from the International Finance Corporation and a group of Indian institutional investors. This institution has outstanding management and has had to adapt to the very real problems which exist in India.

The evidence is that if financial institutions generally have a range of options at their disposal they will opt out of housing finance because it is too difficult. Specialist institutions can bring to bear the specialist techniques which in developing countries is so essential. However, they are likely to require the backing of more general financial institutions, and might need to diversify as they develop in order to continue to operate in a viable manner. It is also clear that initially at least, housing finance institutions must raise their funds largely from the wholesale market.

In some developing countries much attention has been given to the concept of a secondary mortgage market and some people are even devoting resources to developing an international secondary market for developing countries. This really is trying to run before one can walk. There is no prospect of international secondary markets in respect of housing finance loans made in developing countries, because there is no way in which the exchange risk can be overcome. This is a case of people believing that because it happens in the USA then it must be alright for everywhere else in the world. This is very far from the case. Possibly, some secondary market techniques could be used in developing countries and certainly most housing finance institutions in developing countries raise their funds predominantly from the wholesale rather than from the retail markets. This does not mean, however, that they should issue mortgage-backed securities. Loans can often achieve exactly the same effect more efficiently.

CONCLUSION

Countries have much to learn from each other in respect of housing finance. Arguably, the most industrialized countries can learn the most. Had the USA looked at what was happening in other countries in the 1960s and 1970s it might have avoided the massive problem which it has had by forcing its institutions to borrow short and lend long. Britain and other European countries are currently benefiting from secondary market techniques which can help reduce the cost of funds for homebuyers. The use made of mortgage insurance is also relevant from country to country.

One rather suspects that developing countries can learn more from each other than they can from industrialized countries. Housing finance in the Third World is a subject in its own right. It needs to be analyzed not primarily by housing finance experts in industrialized countries but rather by experts in the economies and financial structures of developing countries. What is happening in the USA and the

UK is largely irrelevant to what might happen in India or Africa. The unquestioned following of techniques of the industrialized countries is likely to cause considerable damage in developing countries. Fortunately, however, there is an emerging consensus on what needs to be done in respect of both housing and housing finance in developing countries. The consensus sees only a modest role for government with that role being primarily one of facilitation. Self-help and promoting informal methods of housing construction and housing finance are seen as being the way forward.

6
The Housing Finance Agenda in Developing Countries

Bertrand Renaud
Housing Finance Advisor, The World Bank

The discussion of current issues in housing finance in developing countries can be rendered as complex as any analyst could wish. However, three plain facts underlie the housing finance agenda today:

1. Cities are built the way they are financed, and in the residential sector building is often inefficient or poor.

2. The current cost of existing housing finance systems is impossible to maintain in the present financial environment.

3. The potential contribution that housing can make to financial development has yet to be realized in most countries.

To elaborate on these points the following questions are addressed: Why is housing finance increasing in importance today? What is the experience of lenders like the World Bank in the urban sector and in housing finance? What is the structure of the agenda in developing countries?

HOUSING FINANCE INCREASING IN IMPORTANCE

World urbanization is keeping its momentum. The growth of urban areas continues unabated in spite of the economic crisis of the early 1980s and the current uncertain economic environment. This growth

55

is fuelled by demographic forces and on-going changes in domestic economies. The world population is crossing the five billion mark in mid-1987. Within the next 15 years the majority of the world urban population will be in developing countries. Many of the largest cities are in the heavily populated countries of Asia. The fastest growing centers are in the less urbanized countries of Africa where major cities are still doubling their population in less than a decade.

Urban development now takes place in difficult economic conditions. The external environment is characterized by higher and more volatile interest rates, increasing interdependence of financial markets, large international funds movements, heavy foreign debts and weak international trade markets. Internally, many countries are struggling with large public sector debts and weak public sector performance. The domestic financial systems of developing countries have become more vulnerable to international financial shifts over which they have very limited control.

In most developing countries, housing is one of the most important sectors in the economy. Its share in both annual investment and national wealth is very large. It is also one of the sectors most severely affected by the new economic environment. The sensitivity of housing investment to economic uncertainties is caused by both its nature as a very long-lived investment good and by the structure of financial systems that are not responding well to the world's more volatile financial environment. The performance of the housing sector is a major concern given its potential magnitude and its direct and indirect impact on domestic savings. Imbalances within the housing finance sector are currently growing almost everywhere and investment in housing is severely disrupted in heavily indebted countries. Therefore, policy-makers also face social problems which will not solve themselves on their own: housing affordability and the capacity to borrow are falling; the provision of basic shelter for the poor is deteriorating.

There is a need to evaluate how and where housing finance policies can improve urban policies. Sound housing finance policies and operations can play a critical role in improving the quality of investment in the urban sector. Correctly structured, they can reduce the need to resort to the patchwork of policy responses which have been well intentioned but have proved quite ineffective in addressing housing and urban problems such as rent control laws, land ceilings acts, and unaffordable standards. Clearly, more efficient finance systems are essential to avoid the continued deterioration of urban conditions.

Better housing finance will improve housing and urban invest-ment. It can also improve fiscal and financial policies considerably. In the unstable economies of heavily indebted middle-income countries, housing finance is a part of the restructuring strategies essential to restoring growth. In other countries, housing can con-tribute significantly to expanding the process of financial inter-mediation and the market deepening which has been halted by infla-tion. In the present world financial environment, the short-term and long-term macro-economic benefits of better housing finance policy can be quite large. On the negative side, the costs of maintaining ineffective housing finance systems can often be extremely onerous.

In most developing countries, financial institutions provide only a small portion of the supply of mortgage credit. Most credit for housing investment is provided through informal, barter-like agree-ments or, for development and construction, through high-cost indigenous lenders. In spite of its smaller absolute size, the formal housing finance system is the important one in policy for two reasons. First, the size and dynamism of the informal sector provides a measure of the failure of the formal financial system to meet household demands. The informal financing of housing is often a very creative one, but it is caused by the regulatory constraints placed on the emergence of housing finance institutions. To focus on formal housing finance is to address the constraints which impede efficient financial intermediation. Second, housing is a large component of economic activity. In developing countries it has a rising share of investment and GDP; in developed countries it has an enormous share of the fixed-capital stock (in the US over 50 per cent). When that sector is not integrated in the financial system the economy suffers efficiency losses; the financial system does not provide the information about relative prices and investments that it could. For these reasons, the formal housing finance system must be a continuing concern of policy-makers.

EXPERIENCE OF LENDERS LIKE THE WORLD BANK
IN HOUSING FINANCE

Housing finance operations are a major new item on the World Bank's agenda, and loans with a significant housing finance component total

Table 1.   Housing Finance Loans by the World Bank[1]

Country	Loan Name	Amount	Year
Morocco	First Housing Loan to Credit Immobilier et Hotelier (CIH)	59.0	FY 83
Zimbabwe	Urban Development I	43.0	FY 84
Senegal	Technical Assistance for Urban Management and Rehabilitation	6.0	FY 84
Malawi	Urban Development I (Follow-up of Structural Adjustment Loan I of 1981)	15.0	FY 85
Chile	First Public Sector Housing Project	80.0	FY 85
Nigeria	Second Urban Development Project	85.0	FY 85
Portugal	Housing Finance Project (cofinanced)	25.0	FY 86
Mexico	Low Income Housing Project (FONHAPO)	150.0	FY 86
Indonesia	Housing Sector Loan (Bank Tabungan Negara)	275.0	FY 86
Korea	Housing Finance Sector Loan	150.0	FY 87
Thailand	Urban Development III	21.0	FY 87
Ivory Coast	Urban Development III	126.0	FY 87
Mexico	Housing Finance (FOVI/Commercial Banks)	300.0	FY 88
Philippines	Housing Sector Loan	175.0	FY 89
Ecuador	Second Low Income Housing Loan	60.0	FY 88
India	Housing Finance — HDFC	100.0	FY 88
Argentina	Housing Finance Loan	200.0	FY 89
Tunisia	Housing Finance and Cadastre Restructuring	80.0	FY 89

1. A summary inventory of past loans and loans in preparation which in 1987 had a significant housing finance component (in millions US$).
Note: This list of loans adds up to $2.0 billion. Larger amounts for the India and Tunisia loans are under consideration. A number of operations have been identified and are at various stages of preparation in Latin America, Africa and Asia. Sector work on housing finance issues is also in progress or in preparation in countries other than those listed above (Pakistan, Bangladesh, Nigeria, Hungary, China, Algeria, Turkey).

$2.0 billion (Table 1). There is a rising demand for housing finance loans by member countries, and the average loan size is also increasing. New housing operations concentrate on financial institutions, mortgage instruments and domestic financial resource mobilization. They differ considerably from earlier urban loans as housing policy has moved toward a greater emphasis on financial issues. At the same time, the changes in the financial environment have created serious fiscal and financial problems which require a reordering of financial priorities. At this new crossroad, Bank operations are increasingly supporting government efforts to redirect activities away from the direct provision of housing in order to concentrate on the lack of formal sector mortgage finance. The other major concern is the supply of serviced urban land.

Early urban projects emphasized appropriate urban design and better standards. They involved considerable efforts to upgrade low-income areas. This work broadened almost immediately to the improvement of urban agencies, but it remained primarily project-influenced. Meanwhile, the international consensus on housing policies has increased and the agenda has shifted, partly from the demonstration effect of World Bank-sponsored urban projects. Objectives such as low-cost standards or cost recovery which were often controversial are now accepted, if not always achieved. The do's and don'ts of financing low-income housing programs remain important. However, financing low-income housing or welfare programs is not the exclusive, not even the primary, purpose of nationwide housing finance systems.

There is a need for a more comprehensive evaluation of how significant housing finance has become for developing countries. The interactions between the financial system and the implicit fiscal policies hidden in housing credit subsidies are a critical part of a review of housing finance. The goal is a financial system which can serve housing efficiently as a major investment sector, especially in rapidly urbanizing countries. It is also a housing finance system that must not become a mechanism providing large and untargeted resource transfers whenever macro-economic conditions deteriorate. Such a system will free scarce public resources to concentrate on public investments and the social needs of the poorest.

In the unstable economies of heavily indebted countries, the macro-economic impacts of public housing programs is another important cause for concern. The effects of ill-designed housing policies are spilling over from the housing sector into the rest of the economy and public programs must be corrected.   The impact of these programs on the poor through rapidly rising real housing rents can also be very large and severe.   In middle-income countries in particular, housing finance is often a significant part of structural adjustment needs.   Disruptive housing finance programs affect strategic areas of the economy including: financial resource mobiliza-tion due to negative returns on savings; the efficiency of government expenditures due to low productivity in the public sector and their level due to large contingent liabilities; the inadequate sharing of financial risks between lenders and borrowers under severe inflation; and ultimately serious strains on the credibility and the viability of large financial institutions.

STRUCTURE OF THE AGENDA IN DEVELOPING COUNTRIES

Basic Features of Housing Finance

First, let us define the importance of the sector in the economy and examine the characteristics of housing which affect its financing. These same characteristics cause the regulation of housing finance to have broader implications for fiscal and financial policies.

Annual housing investment commonly represents 12-25 per cent of annual fixed capital formation.   This share of investment tends to rise with the level of development.   Housing is also a major component of national wealth and forms a major part of total national tangible assets.   It is also the main form of wealth accumulation for most households: it is a long-lived asset that has a value equal to three to five times annual household income.   For efficient invest-ment it requires long-term financing.   Even though housing can con-stitute very good collateral, the supply of long-term credit is not expanding in most developing countries.   This has been caused in part by government fears that such long-term finance would crowd out other important investments.

Does the increased availability of mortgage credit lead to a decrease in non-residential capital investment?   The historical record for industrial countries does not give evidence of the existence of such a link over the long-term.   Crowding-out does not come from the mere availability of mortgage credit at market rates.   Rather, housing crowds other investments when it receives a preferential financial and/or tax treatment which raises its rate of return compared to other investments. Directed credit in favor of housing (or any other priority sector) may be less of a long-term factor. In mixed economies, when interest rate volatility increases, attempts at directed credit erode through the fungibility of credit. This fungibility also depends on the proportion of mortgage credit which goes to refinance existing housing rather than new construction. Conversely, attempts to redirect investment away from housing succeed mostly in driving urban investment into the informal sector and in reducing the efficiency of this investment.

The historical experience of developed countries is that the extension of mortgage credit at market rates in pace with the general growth of financial markets has been favorable to overall economic development. Policies which emphasize the need for, and the broad benefits of, efficient housing finance should not be confused with preferential investment in the housing sector. The main purpose of housing finance policy is not to better allocate limited, underpriced funds to public housing programs. The question to address is how to realize the potential contribution that housing can make to financial development: what kind of financial services will both improve the efficiency of investment in housing and increase households savings in financial form. From a social point of view, a housing finance system which meets the needs of the majority of families who can pay without implicit assistance will allow scarce public resources to be redirected to the most pressing needs of the very poor, that is, health, nutrition, education and basic urban services such as water, sanitation and transport.

Heritage of Earlier Policies

To understand present problems it is important to retrace the impact on housing of earlier economic planning and financial policies in developing countries. The Second World War was followed by a long period of growth with low interest rates and price stability. In the beginning of this period, the world economy and capital markets were not well integrated. In order to achieve the goal of accelerated development, most developing countries engaged in various forms of directed credit. Such management of the financial sector could be viable as long as external trade was small, inflation insignificant and interest rates stable. These policies of directed credit have included a deliberate effort to preempt household savings for priority sectors and the rigid regulation of institutional housing finance. Unfortunately, strong directed credit conflicts with the integration of the world economy, technological change, the global integration of capital markets and inflation followed by volatile interest rates. The various types of financial systems which have evolved from such policies have come under increasing stress.

Main Factors Dictating Current Housing Finance Policies

The housing finance problems that developing countries encounter today are caused much more by (1) the management of their financial systems, (2) their choice of housing policies and (3) their macroeconomic performance than by their level of income and degree of urbanization. This point is frequently missed.

The combination of these factors provides a useful way to compare individual countries and to structure the policy agenda. Regarding financial policies, some countries — especially in East and South Asia — have traditionally repressed financial systems which they are now attempting to modernize and liberalize. Others, especially in Latin America, are facing financial instability because of earlier policy errors and want to return to stability. Regarding housing, some countries attempt to suppress institutional housing finance through rigid regulations. Others heavily subsidize housing credit for social reasons. Very few countries have achieved the broad benefits of encouraging households to bid for funds at full price for their investments. Finally, many developing economies are resuming their growth at a lower rate following the shocks of the early 1980s, but some are unfortunately suffering from a pattern of secular decline.

For the purposes of this paper, the problems can be regrouped with reference to two operationally important categories of countries. First, there are the stable but financially repressed systems using strong directed credit where housing can play an important role in the development of the financial system. However, to mobilize more savings through the financial system, savers must be properly compensated. Mortgage rates must reflect the true cost of capital to achieve this objective. Second, in unstable, high-inflation countries, housing finance policy must first limit the very large and poorly targeted resource transfers caused by inadequate mortgage pricing. In such countries the lack of indexation, or indices linked to the wrong bases, create large implicit subsidies and major contingent liabilities for government.

Agenda for Stable but Historically Repressed Financial Systems

In the first group of stable countries, housing finance can play a significant role in expanding financial intermediation and contribute to financial deepening.   However, the traditionally rigid regulation of housing finance has become inconsistent with the uncertain economic and financial environment.   The exclusive emphasis of many economic and urban plans on quantitative targeting and direct control of credit flows is increasingly problematic.   Efficient housing investment requires much greater attention to the structure and adaptability of incentives in housing finance.   Flexibility in housing finance is needed because segmented financial systems are extremely vulnerable in increasingly interdependent and volatile financial markets.   Existing institutions may be changed, or new ones allowed, to permit prices and interest rates which reflect correctly the changing cost of resources and shifting economic conditions.   The necessary steps are likely to include changes in the regulatory framework and new mortgage instruments.

Agenda in Heavily Indebted and Unstable Economies

Housing policies in the second group, the heavily indebted and unstable economies, need to be quite different until the return of stability.   Poorly designed social housing programs can have a very large macro-economic impact on the performance to these disrupted economies.   Housing can have a major role to play in strategies to restructure the economies of Latin America and Turkey, for example. Given the constraints imposed by falling real incomes and the insolvency of most public financial institutions, the emphasis must be placed on possible ways to restructure existing housing finance programs in order to move these economies back to some degree of normalcy.   Under present circumstances housing finance policies can and must contribute to restoring or increasing public confidence and government credibility.   The main way this can be done is by containing the open-ended public subsidies, implicit resource transfers and contingent liabilities linked to the design of mortgage instruments.

CONCLUSION

The economic costs of existing housing finance systems are already high and becoming increasingly impossible to maintain in the new

financial environment.   Too often, heavily subsidized public programs
are not meeting their social objectives.   Still, they impose significant
financial and fiscal burdens on the rest of the economy.   Meanwhile,
the contribution that housing finance can make to overall financial
development and better resource allocation is not taking place
adequately.   The efficiency of investment in the urban sector is lower
because the supply of mortgage credit is not evolving properly.
Instead of flowing through the financial systems, a high proportion of
domestic savings prefers to remain in the informal sector; this limits
the efficiency of investment in the entire economy.

Should we be optimistic about the future? I believe so.   Many
countries are taking a fresh look at their traditional ways of managing housing finance.   The economic pressures for change may be
heavy, but the changing technology and their rising skill level are
such that opportunities for improvements are not likely to be missed.
Changes will not be immediately visible, but in five years or so they
will be there.

Acknowledgement

The views presented here are the author's own and should not be
attributed to the World Bank nor any of its affiliated organizations.

7

Improving Urban Infrastructure Planning and Investment: The World Bank Experience

John M. Courtney
Senior Urban Planner, The World Bank

Urban development and the overall economic development of a country are interconnected. Cities make vital contributions to economic growth. Although the urban population of developing countries usually accounts for between 20 and 40 per cent of the population, well over 50 per cent of the gross national product is produced in cities and towns. By the year 2000, the average is projected to rise to 67 per cent, and 80 per cent of the annual increment to GNP is expected to be from urban areas. If a country's cities and towns are inefficient, then the economy is inefficient, and economic recovery and longer-term development are limited.

Overcoming weaknesses in, and finding new solutions to, urban infrastructure planning and investment have been identified as important objectives by governments and international urban lending agencies. When the World Bank first began working on urban problems in the 1970s, project loans concentrated mainly on construction of facilities and investments, particularly for shelter and transport. Now the Bank has shifted to strengthening institutions and supporting efforts to improve urban planning and management, investment programs, municipal finance, maintenance and operation of urban infrastructure and clearer working relations between central and local governments.

This paper briefly illustrates and comments on some examples of the new approaches that have emerged over the last few years.

The case studies cited are part of the World Bank action planning review of project work and represent a "process" approach, which is a significant change from the earlier "project" approach of traditional World Bank lending. They reflect a shift toward the urban planner's approach to analyzing the city as an entity and looking at the sectoral linkages and trade-offs as a way of formulating an investment strategy. The case studies are of Indonesia, Philippines, Turkey, Korea, Mexico and Nigeria. They are drawn from the six World Bank regions and exhibit a range of country-specific conditions reflecting different stages of institutional development and political persuasion. The common themes are an emphasis on building intermediary institutional capacity for large-scale service and infrastructure. Each project involves improved planning processes and stresses longer-term sustainability and replication on a national scale. Details are given in the annexes.

THE CASE STUDIES

Briefly, the case studies reviewed are as follows:

Indonesia: Integrated Urban Infrastructure Development Planning (IUIDP)

The basic approach is for local government to take the lead, with national sectoral agencies providing advice, in preparing: (1) a medium-term development plan (five to seven years) which identifies needs and priorities spatially and by sector; (2) an investment and operation and maintenance organization and management plan and program by sector with a multi-year budget for both national and local components; and (3) a financing plan and local revenue development strategy.

Philippines: Manila — Capital Investment Folio Process (CIF)

The Philippines has embarked on an innovative approach to urban planning and investment designed to bring realism to sectoral investment programs. The CIF process serves to generate inter-agency consensus on what should happen and thereby influence policy and line agencies of central and local government. It also is intended to function as an operational mechanism for metropolitan management.

Turkey: Cukurova Urban Development Project

The project is formulated within the framework of the government's decentralization policy. The main objectives are to: (1) assist five municipalities to overcome service deficiencies and manage urban growth through the financing of urban infrastructure; and (2) to introduce policies and institutional arrangements for investment planning and implementation, cost recovery, financial management, and staff development which would be suitable for replication in other Turkish cities.

Korea: Pusan Urban Management Project

The project focus is on the improvement of Pusan's urban management and finances. Project objectives are to: (1) improve the organization and coordination of the city; (2) strengthen and reinforce sub-project selection and investment planning; (3) establish a transportation management system to coordinate and optimize transportation planning and investment; (4) strengthen the financial, planning and managerial systems of the city; and (5) support priority investments and the sound development of the city.

Mexico: Municipal Strengthening Project

The project supports the government's objectives to improve the utilization of human and financial resources and strengthen institutions, particularly at the municipal level, and to increase the delivery of affordable and efficient urban services. Within these overall program objectives, the project would: (1) strengthen the municipal organizational, administrative and financial management systems and performance; (2) strengthen federal and state agencies in the area of municipal development promotion through provision of financial resources and effective training programs; and (3) assist municipalities in expanding and upgrading infrastructure.

Nigeria: Infrastructure Development Fund (IDF) Project

The primary objective of this project is to establish an urban infrastructure wholesaling mechanism using merchant banks to appraise, supervise, and cofinance state urban infrastructure projects. These projects would assist the states to manage, maintain, and consolidate existing urban infrastructure and services and to improve financial management and resource mobilization.

REVIEW COMMENTS AND RECOMMENDATIONS

The infrastructure-deficient conditions of the urban areas of the Third World highlight the need for responsive infrastructure planning and investment processes. Departures are needed from the more traditional planning approaches used for these cities in the past. A number of promising approaches are currently underway and lessons may be extracted from them. The challenge remains for the urban practitioner to continue to develop and refine these approaches to meet special conditions in each location. Based on the review to date of the six World Bank-financed case studies, the following approaches are recommended as essential to the effective future development and implementation of urban infrastructure planning and investment strategies and the management of urban areas.

1.   Use an improved strategic planning process to increase spatial, institutional, and financial coordination of investment among various sectors.

2.   Develop and adapt the planning processes so that they are responsive to the design and implementation of affordable urban infrastructure investments, particularly for the urban poor, and establish financial mechanisms that would permit the replicability of urban investment.

3.   Involve the private sector, formal and informal, in providing shelter and urban services to low-income groups at affordable prices.

4.   Delegate authority and accountability for investments from the central to the local level in both political and financial terms. This is essential to the effective future development and implementation of planning strategies and the management of urban areas.

5.   Work within local institutions, rather than remove them on the grounds of efficiency, and work within the constraints of local cultural values.

6.   Include support for development institutions, their administrative mechanisms, and the education and training of their staff.

Acknowledgement

The views and interpretations in this paper are those of the author and should not be attributed to the World Bank, to its affiliated organizations, or to any individual acting on their behalf.

BIBLIOGRAPHY

Courtney, John M. 1985, "Urban Development, Shelter and Basic Needs in the Third World," Paper presented at the Commonwealth Association Conference, Jamaica, West Indies, June 1985.

Courtney, John M., 1986, "Urban Project Implementation: Some Insights from the Practitioner," Paper presented at the 28th Annual Conference, Association of Collegiate Schools of Planning, Milwaukee, October 1986.

Lea, John P. and Courtney, John M., 1985, "Cities in Conflict: Studies in the Planning and Management of Asian Cities," The World Bank, Washington, DC.

Linn, Johannes, F., 1983, "Cities in the Developing World: Policies for Equitable and Efficient Growth," Oxford University Press, New York, NY.

Loh, Ping Cheung, 1986, "Policy and Programs for Urban Shelter Lending: The Next Decade," speech presented at the Second International Shelter Conference, Vienna, Austria, September 1986.

Taylor, John and Williams, David, eds., 1982, "Urban Planning Practice in Developing Countries," Pergamon Press, Oxford.

The Urban Edge, 1987, "Setting Priorities in a Large City: Financial Management in Pusan," Vol. 11, No. 3:4-5, April 1987.

Water Supply and Urban Development Department, 1986, "Urbanization in the Developing Countries: Issues and Priorities," Water Supply and Urban Development Department, The World Bank, Washington, DC: October 1986.

Williams, David G., 1984, The role of International Agencies: The World Bank, *in*: "Low-Income Housing in the Developing World," G. K. Payne, ed., John Wiley:173-185, Chichester.

Wright, Albert M., and Courtney, John M., 1986, "Strategic Sanitation Planning," Water Supply and Urban Development Department, The World Bank, Washington, DC.

APPENDIX: CASE STUDIES

Indonesia: Integrated Urban Infrastructure Development Planning (IUIDP)

In order to improve the coherence and impact of the urban infrastructure program and to develop better capabilities in planning, budgeting, management and coordination at the local level, the government of Indonesia has initiated an improved planning and evaluation process for urban investments known as Integrated Urban Infrastructure Development Programming (IUIDP). The basic approach is that the local government would take the lead, with sectoral agencies providing advice, in preparing: (1) a medium-term development plan (five to seven years) which identifies needs and priorities spatially and by sector; (2) an investment and organization and methods (O & M) plan and program by sector with a multi-year budget for both "national" and "local" components; (3) a financing plan and local revenue development strategy; and (4) a yearly budget submission.

The development plan and yearly budget would be submitted to the provincial government for allocation of grant and loan funds. In addition to this review, the provincial government would also provide guidance on IUIDP procedures and assistance to local government as needed. The central government role would be limited to: (1) setting standards; (2) providing financing (grant and loan) mechanisms; (3) providing guidance on the IUIDP planning process for regional governments and technical assistance when necessary; (4) appraising metro and large-city IUIDP programs; and (5) training local staff in planning, implementation, and O & M procedures.

The IUIDP process is a much-needed and overdue reform in Indonesia. Managed properly, it could offer: (1) integrated planning of infrastructure based on rational technical criteria and standards but tailored to a city's economic and social requirements; (2) improved coordination between central, provincial, and local government investments in urban areas; (3) improved costing and budgeting for O & M requirements; (4) an incentive for local revenue generation; (5) coherent city-level plans, programs, and project priorities which can be appraised by provincial and central government with minimum changes or duplication, and (5) links to a Regional Government Loan Fund (FGLF), and the recently-approved First Urban Sector Loan for Indonesia.

Philippines: Manila — The Capital Investment Folio (CIF) Process

Recognizing traditional urban planning deficiencies, the government of the Philippines in 1978 embarked on an innovative approach to urban planning which was designed to bring realism and relevance to sectoral investment programs. The CIF was initiated as the Metro Manila Financing and Delivery Services Project (MMETROFINDS) within the Ministry of Public Works and subsequently transferred to the Metro Manila Commission (MMC), where it is now the responsibility of the Office of the Commissioner for Planning. The approach is designed to remedy the basic problems of traditional urban planning. It focuses on three major components of the planning process and on their interlinkages. The main features of the approach may be summarized as follows:

1.   The development of a Framework Plan for Metro Manila recognized by all major agencies, which provides the basis for policies and projects to be identified and evaluated.

2.   Identifying proposed expenditure on all public sector projects and programs in Metro Manila; and separately estimating likely available funding for such projects.

3.   Developing an evaluation procedure to establish priorities.   In practice the demand for funds always considerably exceeds availability so that projects must be cut back until demand and availability are in balance.

The CIF is in effect the financial arm of the Framework Plan for Metro Manila, the group of high priority public sector projects that contribute to the government's objectives. The basis of this new approach is to generate interagency consensus of what should happen and thereby influence the policy and line agencies of central government and the local governments. Although the CIF process has been delayed due to recent political changes in the Philippines, its operational mechanism for metropolitan management is probably unique in its present form and is generating considerable international interest, as well as interest within the Philippines.

Turkey: Cukurova Urban Development Project

The Cukurova region was selected for this project because it exemplifies the problems of rapid urban growth and service deficiencies, and because future growth in this important area will be jeopardized if services are not provided more efficiently. Urban service deficiencies hamper equity and growth prospects.

Almost half of the population of the five largest Cukurova cities live in *gecekondus* (urban slums), with higher proportions in the larger centers (Adana, 58 per cent; Mersin, 47 per cent; Tarsus, 40 per cent; Iskenderun, 30 per cent). Some *gecekondus* are built on public land but most building is not on private land subdivided and sold without planning approval. The majority of these settlements are served, formally or informally, with water and electricity, some with roads, but few with sewers. Drainage is also a problem in large parts of these cities.

Lack of basic infrastructures — water, electricity, and telecommunications — increasingly constrain development of light industry, although major increases in power and telecommunications are planned or underway. The proposed project is formulated within the framework of the government's decentralization policy. The main objectives are to:

1.   Assist the municipalities of Adana, Mersin, Tarsus, Iskenderun, and Ceyhan to overcome service of deficiencies and manage urban growth through the financing of urban infrastructure.

2.   Introduce in the five project municipalities policies and institutional arrangements for investment planning and implementation, cost recovery, financial management, and staff development which would be suitable for replication in other Turkish cities.

3.   Build up in Iller Bank a capacity for appraisal and monitoring of municipal infrastructure projects and investment programs.

The project would finance the 1987-92 section of the project municipalities' investment programs in basic engineering services. These investment programs were developed during project preparation and would be updated and revised annually to respond to changing demands and resources. The investment programs were designed to address service needs in a systematic coordinated manner with special emphasis given to present and future resource constraints and service deficiencies in low-income areas.

Korea: Pusan Urban Management Project

With a population of 3.5 million, Pusan is the second largest city in Korea and has 15 per cent of the country's total urban population. It is an important industrial, commercial, and educational center and is the largest port in Korea. During the last five years, Pusan's major investment has been 100 km of subway lines at a cost of $3.5 billion. Financial constraints have forced the city to lower its sights and re-examine the adequacy of its investment and financial planning. The subway and other large investments, mainly for highways, water supply and pollution treatment, averaged $400 million per annum between 1983-86. However, these investments lacked adequate prioritization, and were not undertaken in the context of a well-formulated strategy linking long-term plans and programs with available resources. In particular, the subway investments forced a postponement of smaller but higher priority investments which would allow completion and full utilization of existing infrastructure. A major cause of these problems is the city's institutional arrangements which hinder urban management.

The project focus is on the improvement of the city's urban management and finances. Project objectives are to: (1) improve the organization and coordination of the city; (2) strengthen and reinforce subproject selection and investment planning; (3) establish a Transportation Management System (TSM) to coordinate and optimize transportation planning and investment which is the city's most critical problem; (4) strengthen the financial, planning and managerial systems of the city; and (5) support priority investments and the sound development of the city.

The proposed Bank loan of $50 million would finance 30 per cent of a three-year (1987-89) priority investment ($165 million) by the Pusan city government and would support important institutional development improvements. The project investment forms part of the city's investment plan, with emphasis on small complementary investments to complete the benefits of existing infrastructure, they include: (1) construction and improvements of priority roads (expansion, rehabilitation, paving) and a comprehensive program of low-cost transportation improvements; (2) drainage, flood protection and sewerage works; (3) embankment and steep slope protection; and (4) improvement and expansion of city services (markets, landfill, fire engine equipment, community facilities). Most of these are in low-income areas. The improved streets would allow the provision of needed urban services (water supply, sewerage, solid waste collection, postal services, and so on).

The city's new investment plan is the first step in an effort to rationalize the selection of investment projects, basing overall investment priorities on standard economic, technical, and financial criteria. Using such an approach, it became clear that the city would achieve the greatest immediate benefits by shifting from large and highly visible new works to a range of smaller investments that permit the full use of existing infrastructure. For example, measures to integrate subway and bus systems — including new bus routes, road improvements, and transit terminals — will enable the city to get greater benefits from the subway system at relatively little additional cost.

Under the project, the city would establish an Investment Mechanism to improve project selection, set adequate approval criteria for all subsectors, and prepare and update long-term investment plans. The unit would review all investments over $5 million and set priorities based on economic, technical, and financial criteria.

Mexico: Municipal Strengthening Project

Mexico's outstanding spatial problems are: (1) the heavy concentration of economic activity, wealth, and population in Mexico City; (2) the lack of integration between urban and rural areas; and (3) unbalanced interregional development. Approximately 63 per cent of the country's 75 million people live in urban centers of more than 2500 inhabitants, while about 26 per cent of the population live in the three main metropolitan areas, Mexico City, Guadalajara, and Monterrey.

Providing basic municipal services and improving the social well-being of all Mexicans has preoccupied successive administrations. During the 1970s and early 1980s, the population with access to piped water increased from 40 to 66 per cent and connections to sanitation systems increased from 29 to 43 per cent. However, the recent economic climate has led to a significant rise in service deficits nationwide and a deterioration in existing infrastructure through poor maintenance and lack of training.

The proposed project emphasizes a "process" orientation that would represent a shift away from traditional single project analysis to an institutional appraisal focussed on the municipality as an important agent of national economic development. The project is designed around an integrated set of mutually supportive activities. After carrying out a detailed institutional and financial appraisal, municipalities would have access to credit for municipal infrastructure subject to presenting subproject appraisals, participating in relevant training programs and implementing required financial measures.

Nigeria: Infrastructure Development (IDF) Project

Nigeria's urban population was estimated in 1986 at over 30 million, more than 30 per cent of the total population, and is expected to rise to 80 million by the year 2000 — one-half of Nigeria's total population. Investments in urban areas have not kept pace with growth and were often poorly selected and executed, thereby hampering the cities' ability to provide an efficient operating environment for industry and commerce. Recent World Bank sector work identified inadequate infrastructure and financing systems as the most critical urban problems. The IDF project would test a wholesaling mechanism through the financing of initial Bank-approved subprojects in three states and additional merchant bank-appraised state subprojects, and would provide assistance and incentives to states to improve financial management, resource mobilization and project preparation. Long-term efforts would be initiated to involve the Nigerian private sector in the development of urban infrastructure in the states.

BIBLIOGRAPHY

Courtney, John M. 1985. "Urban Development, Shelter and Basic Needs in the Third World", paper presented at the Commonwealth Association Conference, Jamaica, West Indies, June 1985.

Courtney, John M. 1986. "Urban Project Implementation: Some Insights from the Practitioner", paper presented at the 28th Annual Conference, Association of Collegiate Schools of Planning, Milwaukee, October 1986.

Lea, John P. and Courtney, John M. 1985. *Cities in Conflict: Studies in the Planning and Management of Asian Cities,* Washington, DC: The World Bank.

Linn, Johannes F. 1983. *Cities in the Developing World: Policies for Equitable and Efficient Growth,* New York, NY: Oxford University Press.

8

Addressing the Urban Management Challenge: Indonesia's Integrated Urban Infrastructure Development Program

G. Thomas Kingsley
Principal Research Associate, The Urban Institute

INTRODUCTION

We are all familiar with the sizeable gap that exists between developing countries' urban infrastructure requirements and the funds those countries have available to pay for them — a gap that has certainly widened in the troubled economic environment of the 1980s. Financing, however, is not always the binding constraint. Around the world today there are many funded projects that are not moving ahead simply because there is no one available to manage them. More broadly, the lack of trained managers is constraining the expansion of all basic urban government functions in the face of the unparalleled growth of the cities and towns in the developing world. This problem is one of the major barriers to the "replicability" of innovations in urban development (Cohen, 1983) This paper assesses the scope and future of the urban management challenge and describes a promising approach now being implemented in Indonesia: the Integrated Urban Infrastructure Development Program (IUIDP).

THE SCOPE OF THE CHALLENGE

No matter how many times we hear them estimates of urban growth in the developing world are staggering. According to the United Nations (1982), the urban populations of industrial countries are expected to expand from about 0.8 billion in 1980 to 1.2 billion in

76

2025 (an annual growth rate of 0.9 per cent).   In contrast, developing country urban populations are expected to grow from 1.0 billion to 3.9 billion over that period (a rate of 3.1 per cent annually).   That growth alone (2.9 billion) is 18 times the total US urban population in 1980, the equivalent of building 360 versions of metropolitan Los Angeles at its current size.

Present capacity to manage urban growth in developing countries is negligible.   Table 1 shows estimates of government employment per thousand population in three settings as of 1980. Developing countries have a respectable 25 central government staff per thousand, compared to 33 in OECD countries and only 13 in the US.   However, they have only four regional and local government employees per thousand population compared to 44 in OECD countries and 62 in the US.

If an effort were made to move developing countries up to the 1980 OECD regional and local government staffing ratio by the year 2025, based on UN estimates of total population, it would require the recruiting and training of 271 million employees just to make up for the 1980 deficit and an additional 152 million to handle the growth. The 423 million total is 32 times the total state and local government employment in the US in 1980.   Given traditional methods of recruitment and training and a traditional concept of the role of local government, the task is massive.   We will have to re-think the functions of urban government and devise significant innovations in their performance if the urban management challenge is to be addressed.

THEMES OF A NEW APPROACH

There are three main avenues of approach to addressing the challenge realistically and these are described below.

Redefining government's role in urban development, giving more responsibility to the private sector and community groups is one approach.   There is a rapidly growing literature on opportunities for private firms and community groups to assume roles that have been traditionally left to government (Roth, 1987).   Examples include incentives to private developers to provide housing for low income groups; contracting with private firms to handle water billing and collections; organizing communities to handle their own solid-waste collection and disposal; delaying the provision of piped water supply

Table 1.   Government Employment per Thousand Population

	USA	OECD Countries	Developing Countries
Central government	13	33	25
Regional and local	62	44	4
Total	75	77	29

Source: USA data calculated from 1980 Census of Population and Bureau of Labor Statistics reports.   Other estimates based on surveys of 16 OECD countries and 31 developing countries as presented in Ozgediz (1983).

where household pumps and ground water are reasonably adequate to meet current needs.

Improving the internal efficiency of local government is another approach.   Applications of more systematic methods of doing the government's business lie behind most potential innovations in this area and the opportunities are much enhanced by the accelerated application of microcomputers throughout the world (see Cochrane 1983; Kubr and Wallace, 1983; and Panel on the Use of Microcomputers for Developing Countries, 1986).   Another important opportunity is the expanded use of paraprofessionals in many activities.   This opportunity also depends, of course, on systems applications which permit careful delineation of tasks that can be performed competently by individuals without advanced education.

Realistic training with more emphasis on management is a third approach.   The most important avenue may be combining systems approaches and training on the job.   Without leaving the work place, an employee is given a new technique and learns how to apply it directly in handling a problem he is going to have to deal with after the trainers depart.   In addition, there must be growing emphasis on training people who will see themselves as managers playing innovative and entrepreneurial roles not normally associated with the term "administrator".

Much past literature on public administration in developing countries stressed the glacial pace of change in established institutions and the dangers inherent in imposing western management techniques on other cultures.   However, the themes noted above can make a considerable difference in a reasonable period of time. Partially, this stems from the fact that increasingly severe resource constraints are forcing developing country leadership to experiment more boldly with new ways of doing things.   I also take heart from the rapid acceptance of microcomputers in so much of the world. Although their software forces systematic ways of looking at problems and solutions that may be alien to traditional modes, software suppliers cannot keep their shelves full enough.   A third cause for my optimism is the strong recognition of the importance of the management challenge by international donors (World Bank, 1986).

INDONESIA: BACKGROUND AND URBAN STRATEGY

With 160 million people, Indonesia is the world's fifth largest country.   it stretches about 3000 miles from east to west (about as broad as the US).   Its geography is dominated by five land masses — Java, Sumatra, Kalimantan (southern Borneo), Sulawesi, and Irian Jaya (western New Guinea) — but it incorporates over 13,000 inhabited islands in all.   In part due to a successful family planning program, Indonesia's overall population growth rate will be lower over the 1980-2000 period (1.8 per cent) than it was during the 1970s (2.3 per cent) but its labor force growth rate will be higher (2.6 per cent compared with 2.1 per cent).   To absorb that growth, the nation will have to create an average of 1.8 million new jobs annually through the end of the century — almost one-third above the actual job creation rate of the 1970s.

Although petroleum resources have played the most visible role over the past two decades, Indonesia's economy is still dominated by agriculture.   Structural change is rapid, however.   Manufacturing, services and other sectors that tend to locate mostly in the cities are significantly expanding their shares of the total work force.

Concerned about the potential for formidable urban growth, Indonesia initiated a National Urban Development Strategy project in 1982 with assistance from the United Nations Center for Human Settlements (UNCHS).   The project team had to address Indonesia's longstanding worries about spatial imbalance.   As late as 1980, one island, Java, housed 62 per cent of the nation's population on only 6 per cent of its land area (Java's density was 1700 persons per square mile compared to an average of 80 per square mile in the rest of the country). There was also a concern about the dominance of the capital city, Jakarta, which had grown by 4 per cent annually over the 1970s reaching a population of 6.4 million in 1980.

The team's analysis indicated that trends toward decentral-ization were already under way. While the islands outside of Java accounted for only 35 per cent of Indonesia's 1961 population, they accounted for 41 per cent of its 1961-71 growth and 46 per cent of its 1971-80 growth. Metropolitan Jakarta actually represents a smaller share of the nation's total urban population (19.6 per cent) than do the primate cities of many other countries and that share has remained relatively constant. City growth rates were not simply a function of urban size; cities in the 100,000 to one million range had actually been experiencing the most rapid growth on average. Growth was particularly rapid in port cities in that size class outside of Java and well distributed throughout the archipelago (Kingsley, Stolte, and Gardiner, 1985).

With this background, the resulting strategy did not attempt to dramatically alter recent trends (National Urban Development Strategy Project, 1985). Rather, it suggested that the following effec-tive macro-sectoral policies should be relied upon to enhance existing incentives toward decentralization: (1) diversification of agriculture and manufacturing and the growth of sectors that would be more labor intensive, more export oriented, and generally less likely to induce a concentrated spatial pattern; (2) reduction of protection and taking other steps to improve the competitiveness of older import substitution industries, many of which are based in and around Jakarta; and (3) taking other steps to reduce current location biases favoring Jakarta and other large cities, such as streamlining business regulations, liberalizing and decentralizing the provision of credit and expanding local revenue generation and cost recovery (discussions of the merits of similar approaches can be found in Richardson, 1977; Renaud, 1981; Linn, 1983; and Hamer, 1985).

COSTING OUT URBAN INFRASTRUCTURE

The project put great emphasis on linking strategy to implemen-tation. Accordingly it built a mechanism for testing alternate infra-structure programs, described in a fair amount of detail for each of the nation's 508 most important cities and towns.

Data from various sources were compiled to create an inventory of existing infrastructure conditions for each city; for example, water production capacity per capita, percentage of households connected to piped water systems, lineal feet of paved roads, and so on. Various standards were applied to support analysis of comparative deficits.

Alternative growth scenarios were designed based on differing assumptions about the economic environment, government policy, and spatial incentives. Each contained explicit estimates of population growth and other variables for each city. In all scenarios population growth was distributed primarily across existing cities, avoiding the comparative high costs of self-contained new towns and, to assure reasonableness, estimate ranges were constrained by recent trends.

A computer-costing model was built which yielded explicit estimates of the costs of infrastructure needed to make up current deficits and to provide for growth in each scenario. Model estimates permitted explicit variation of standards and targets by time period for each infrastructure sector. Outputs, in effect, resembled trial capital budgets for each city (National Urban Development Strategy Project/PADCO, 1985).

The results of the final mode, summed across cities, are given in Table 2. The most important conclusions drawn from the analysis were as follows.

Differences in spatial patterns had comparatively little effect on total costs, but costs for individual infrastructure components did vary significantly across scenarios. For example, the most spatially concentrated scenario required higher costs for urban services (water supply, drainage, urban roads, and so on), while the most dispersed scenario showed higher costs for rural road networks, transmigration, and so on. Overall, however, these variations balanced out. Total 1985-2000 public sector investment requirements for all scenarios fell in the Rupiah 70-75 trillion range (at 1980 prices — the equivalent of US $70-75 billion). Given probable estimating error, the small differences in reported totals were immaterial for policy.

The most important factor in cost determination was the setting of infrastructure standards and land development techniques. The outcomes in Table 2 are but the last in a series of trials using different standards and development approaches. Financial studies suggested that about Rupiah 70 trillion was likely to be available nationally for infrastructure over the period, an average of about 25 per cent of gross domestic investment).   The trials made it clear that, to be affordable, Indonesia's urban infrastructure program would have to insist on quite modest standards, low-cost technologies, much expanded cost recovery, and more extensive reliance on community groups and the private sector in the development process.

RECOGNIZING THE MANAGEMENT CHALLENGE

By its detailed accounting of the work to be done, this analysis high-lighted the immensity of the task at hand.   Through the mid-1980s, Indonesia had a much better track record than most developing countries in meeting its urban infrastructure needs (Hendropranoto Suselo, 1984).   Over the 1980-2000 period, however, its urban popu-lation is likely to grow by an average of 2.2 million persons per year — twice the 1970s rate.   Clearly, a new institutional environment is required.

In the past, urban infrastructure had been planned by the staffs of central government ministries in Jakarta and provided by their field units.   In the future decentralization would be essential for two reasons.   First, the job was simply too big to be managed effectively from the capital.   Second, the central government no longer had the resources to pay the bill.   Petroleum revenues had accounted for 70 per cent of total central government resources through the early 1980s, but declines in oil prices were seriously eroding the yield from that source.   Also, local taxation in Indonesia has been quite low by international standards.   Much expanded local revenue generation and cost recovery would be mandatory.   Consequently, new legis-lation has been devised to set the longer term framework for decen-tralization and to expand local capacity to raise revenues.

In the meantime, the government has responded directly in three ways.   First, it has created a National Urban Development Coordinating Board.   Although in the past, the lack of coordination between central ministries was often a problem, the central govern-ment would still have to play a critical role in a more decentralized environment.   The new inter-ministerial Board will set overall policies, monitor program accomplishments and problems, and retain important responsibilities in generating and allocating financial support.   Consistent with lessons learned the hard way in other

Table 2. Summary Investment Programs, Indonesia 1985-2000: Alternative Strategy Scenarios — Industrial, Intermediate and Agricultural[1] (Investment in Constant 1980 Rupiah Trillion)

	I Industrial	II Intermediate	III Agricultural
Inter-Urban Infrastructure			
Transportation			
Sea (ports & fleet)	4.20	4.20	3.85
Air (ports & fleet)	2.66	2.94	2.66
Rail and Ferries	3.77	3.61	3.43
Major Inter-Urban roads	10.94	11.01	10.65
Rural service roads	1.51	1.56	1.57
Total	23.08	23.39	22.16
Power	7.16	7.86	6.21
Telecommunications	0.39	0.39	0.32
Rural education facilities	3.64	3.92	4.76
Total Inter-Urban	34.27	35.56	33.45
Intra-Urban Infrastructure			
Water supply, sanitation & drainage	3.50	3.28	2.40
Roads	6.31	5.27	4.29
Education & health facilities	2.34	2.19	1.80
Power	21.10	16.30	13.15
Telecommunications	3.92	3.50	3.29
Total Intra-Urban	37.17	30.54	23.93
Total Infrastructure	71.44	66.10	58.38
Transmigration (estimate)	4.19	7.42	13.43
Infrastructure & Transmigration	75.63	73.52	71.81

1. Note on alternative strategy scenarios: I = Industrial emphasis; gradual decentralization; II = Intermediate emphasis; rapid decentralization; and III = Agricultural emphasis; accelerating decentralization.

countries, Indonesia resisted the idea of creating a separate ministry for urban development which too often leads to functional confusion and duplication of effort, thereby diminishing focussed attention on urban problems.

Next, the creation of urban governments is required. Decentralization obviously requires the existence of local governments to take on the work. In Indonesia, however, while there were 211 cities with urban populations of 20,000 or more in 1980, only 78 of them had recognized administrative status independent of their rural hinterlands. A process for granting such status and recruiting core staffs has been initiated.

The final move was establishing the Integrated Urban Infrastructure Development Program (IUIDP). Turning over full urban development responsibilities to local governments when those governments have negligible capacity could lead to disaster. The IUIDP program is a process for staged decentralization in which: (1) the central government plays and active role in helping to build local capacity and, at the same time, works with local staff to get on with the business of keeping up with infrastructure requirements. One of the side benefits of this approach has been the coordination of external donor support. Urban sector donors (for example, the World Bank, the United Nations Development Program, the Asian Development Bank) are providing assistance in different provinces, but they are all applying the same IUIDP approach. UNCHS and the World Bank are also providing significant technical assistance to help the government monitor and coordinate the program nationwide.

THE IUIDP - HOW IT WORKS

The IUIDP process was initiated in early 1986. While there are some variations, here are the steps in the "ideal" process. Everything begins with meetings at the province level. Provincial staffs review the outputs of the National Urban Development Strategy for the province, including infrastructure standards, staging and costing assumptions, and adjustments are made as appropriate. Working relationships are established and cities within the province are prioritized for action.

In the next stage, similar meetings are held with staff in the individual cities that have been selected. National strategy outputs are reviewed and adjusted to better reflect local judgements on comparative priorities, across and within sectors, and standards.

A work plan is then developed. This addresses the linking of training with ongoing project work in the locality and often requires consideration of methods of supplementing local government staff and equipment (for example, microcomputers).

Project teams of local staff with technical assistance provided from the center then review and update an existing local land-use plan. Where one is not available, they develop a new "structure plan" indicating broad magnitudes of expected growth by geographic subarea within and around the city and setting the basic pattern for transportation and other infrastructure networks. Unlike traditional "master planning", only a few weeks at most is devoted to this activity in IUIDP.

Infrastructure projects are next identified, scoped and costed. They are then staged and assembled to create a three-year capital budget for the city. Rolling budgets are anticipated thereafter, with plans for years two and three being adjusted, and a plan for year four added, as work under the year one plan is being completed. IUIDP also requires explicit consideration of capital maintenance requirements and arrangements for ongoing operations as a part of capital budgeting. As noted earlier, the computer model from the national strategy project produced what amounted to trial capital budgets for each city. The same model can be extremely helpful in this stage at the city level. It provides a framework for the budget, background analysis of the cost implications of alternative standards, and a capability for quickly making adjustments and testing their effects, and for printing needed reports.

IUIDP then requires the teams to prepare a complete financing plan covering all potential sources of funds. The plan must indicate and allocate expected local revenues, develop a borrowing strategy, and consider amounts that may be needed from the central budget and/or external donors. This step forces the city to think through a strategy for local revenue enhancement over the longer term. Individual city programs so defined are then reviewed and adjusted at the province level and a combined provincial submission is prepared for review at the central level. One of the important tasks at the center is packaging projects in a manner suitable for support by external donors.

CONCLUSIONS

Indonesia's IUIDP program is still at an early stage but so far it has been working pretty much according to plan. While it focusses on infrastructure provision and operations, those activities make up a large part of the job of local governments throughout the world. Thus, IUIDP does address the broader management challenge discussed at the outset.

Its central task, the preparation of a serious local capital budget and financing plan, does much to determine how a local government will actually work in the future. Reviews of its progress by its sponsors should focus on three essential management themes: (1) how well it is allocating responsibilities among the private sector, community groups and government so that development will be manageable over the long term; (2) how well it is promoting internal efficiency within government, using systems approaches to prioritize work and avoid redundancy; and (3) whether the type of on-the-job training being provided, including the introduction of new computer modules to staff as the work proceeds, enhances capacity in a sustainable manner.

Acknowledgement

The author has served as chief technical advisor for Indonesia's National Urban Strategy sponsored by the United Nations Center for Human Settlements. The views expressed are solely the author's own and should not be attributed to any institution.

REFERENCES

Cochrane, Glynn, 1983, "Policies for Strengthening Local Government in Developing Countries," World Bank Staff Working Paper No. 582, The World Bank, Washington, DC.
Cohen, Michael A., 1983, "The Challenge of Replicability: Toward a New Paradigm for Urban Shelter in Developing Countries," in Regional Development Dialogue, Spring.
Hamer, Andres, 1985, "Urbanization Patterns in the Third World: How to Create a Basis for Efficient Growth," Finance and Development, March.
Hendropranoto Suselo, 1984, "Experiences in Providing Urban Services in Secondary Cities in Indonesia," Paper presented to

the 1984 African Conference on the Problems of Urbanization in Africa, Dakar, April.

Kingsley, G., Thomas, 1984, "Urban Strategies and Integrated Urban Development," paper presented at a Workshop of the Directorate General Cipta Karya, Puncak, July. Reprinted in UNCHS Habitat News, Vo. 6, No. 3, December, Nairobi, UNCHS.

Kingsley, G. Thomas, Gardiner, Peter and Stolte, Willem B., 1985, Urban growth and structure in Indonesia, National Urban Development Strategy Project Report T1.6/C3, Directorate General of Human Settlements, Department of Public Works, Government of Indonesia, and United Nations Centre for Human Settlements, Jakarta, August.

Kubr, Milan and Wallace, John, 1983, "Successes and Failures in Meeting the Management Challenge: Strategies and their Implementation, "World Bank Staff Working Paper No. 585, The World Bank, Washington, DC.

Linn, Johannes, F., 1983, "Cities in the Developing World: Policies for Their Equitable and Efficient Growth," A World Bank Research Publication, Oxford University Press, Oxford.

National Urban Development Strategy Project, 1985, National Urban Development Strategy Project Final Report, Report T2.3/3, Directorate General of Human Settlements, Department of Public Works, Government of Indonesia, and United Nations Centre for Human Settlements, Jakarta, September.

National Urban Development Strategy Project/PADCO, 1985, Urban Services Investments for Alternative Strategies, Report T1.8/SC4, Directorate General of Human Settlements, Department of Public Works, Government of Indonesia, and United Nations Centre for Human Settlements, September.

Ozgediz, Selcuk, 1983, Managing the Public Service in Developing Countries: Issues and Prospects, World Bank Staff Working Paper No. 583, The World Bank, Washington, DC.

Panel on the Use of Microcomputers for Developing Countries, 1986. Microcomputers and their Applications for Developing Countries, Westview Press, Boulder and London.

Renaud, Bertrand, 1981, "National Urbanization Policy in Developing Countries," A World Bank Research Publication, Oxford University Press, Oxford.

Richardson, Harry, W., 1977, "City Size and National Spatial Strategies in Developing Countries," World Bank Staff Working Paper No. 252, The World Bank, Washington, DC.

Richardson, Harry W., 1985, "Spatial Strategies, the Settlement Pattern, and Shelter and Services Policies," Paper Prepared for

IYSH Advisory Group Meeting on Shelter, Settlement and Economic Development, United Nations Headquarters, 24-26 April.

Roth, Gabriel, 1987, "The Private Provision of Public Services in Developing Countries," EDI Series in Economic Development, Published for the World Bank by Oxford University Press.

United Nations, "Estimates and Projections of Urban, Rural and City Populations, 1950-2025: The 1982 Assessment," Department of International and Social Affairs, New York, NY.

World Bank, 1986, Urban development: Thinking Big Enough? *in:* The Urban Edge, Vol. 10, No. 2, February, The World Bank, Washington, DC.

Understanding Small Towns in
the Development Process

9

Farm Towns, Mining Towns, and Rural Development in the Potosi Region, Bolivia

Hugh A. Evans
Assistant Professor, School of Urban Planning
University of Southern California

INTRODUCTION

In many less developed countries today, policy makers and analysts are increasingly turning their attention to small towns as a critical element in larger strategies to promote rural development and urban deconcentration. It is thought that small towns can play a significant role in spurring agricultural production, raising productivity, and hence rural incomes. Rising rural incomes in turn spur demand for non-farm goods and services, and are believed to create job opportunities for surplus rural labor in small towns. The prospect of new jobs in smaller towns close to those who seek them supposedly eases the pace of urban growth of the largest cities, and hence the problems associated with such growth (see, for example, Jones, 1986; Kenya, 1986; Mathur, 1982; Rondinelli, 1981, 1983; and UN Center for Human Settlements, 1985).

While such a scenario makes good sense intuitively, there are few empirical studies to support such a thesis. We know little about the conditions under which it takes place, or the reasons why some towns grow, while others do not. Part of the reason is the statistical "invisibility" of small towns in most countries, which impedes detailed analysis (Gibb 1984). Nevertheless, a better understanding of the dynamics of rural-urban growth would help policy makers and planners to determine what kinds of interventions are likely to be more effective in promoting development of smaller settlements, and

91

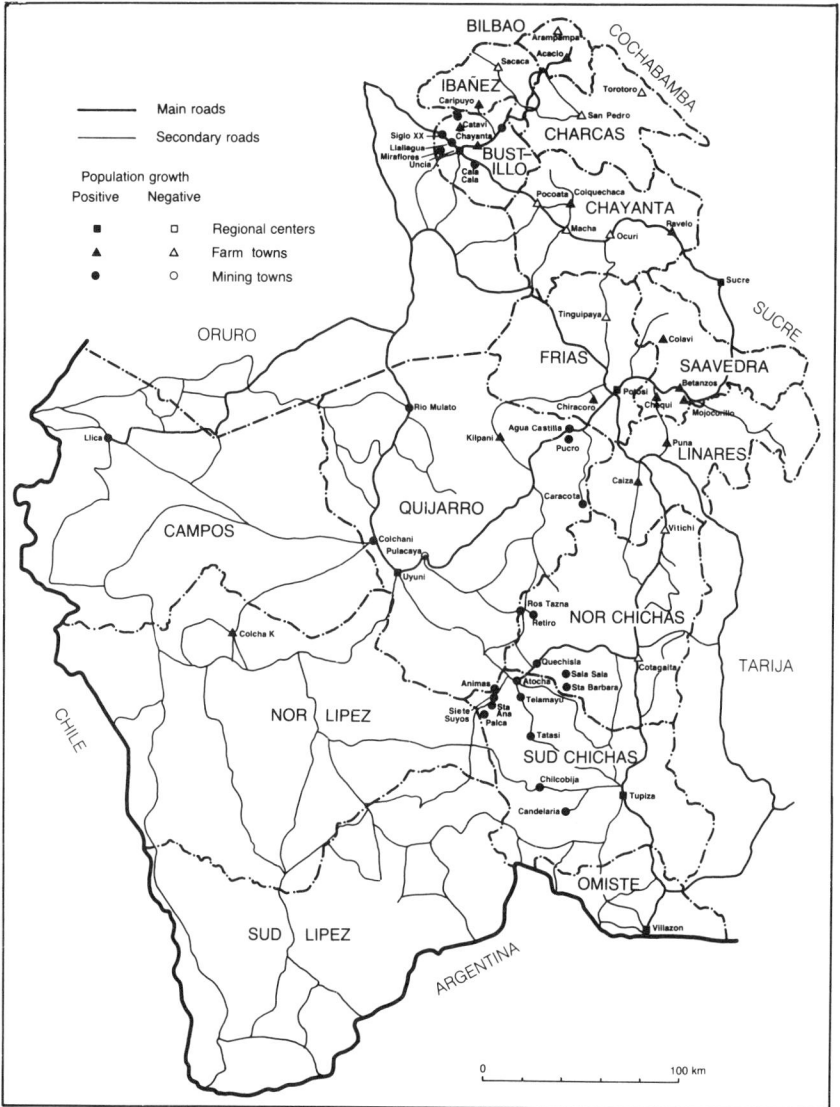

Figure 1. Farm towns and mining towns in the Department of Potosi. (Figure reproduced from "Urban Functions in Rural Development: The Case of the Potosi Region in Bolivia" by Hugh Evans (1982) with the kind permission of USAID, Regional and Rural Development Divisions.)

in which places public investments in infrastructure and services should be located.

This paper attempts to explore some of the links between rural and urban growth in the Potosi region of Bolivia (Figure 1) and the role of small towns in the development process. The data used in this analysis were assembled for the purposes of preparing a regional development plan, while the author was serving as a resident advisor to CORDEPO, the development corporation for the Department of Potosi. Since the data were not collected with the present inquiry in mind, we cannot address the questions as directly as we would like. Instead we have to proceed obliquely, using proxy variables and crude approximations. Nevertheless, the data do provide some clues about the nature of the interactions between basic production activities, and the growth of smaller urban settlements and their rural hinterlands.

THE POTOSI REGION

Despite its enormous earlier wealth, the Department of Potosi is today the most backward region of Bolivia — itself one of the poorest countries in Latin America — and is lagging still further behind. In 1976, per capita income measured by value of production was the lowest in the country at $458 compared to the national average of $637. During the seven-year period 1970-77, output grew at barely 1.3 per cent per annum, compared to the nation's 6.3 per cent and its contributions to GNP fell from 13.5 per cent to 10.3 per cent, a decline unparalleled by any other region (CORDEPO, 1981; Evans, 1982).

The weak local economy has prompted extensive out-migration to other parts of the country, at the rate of 6.4 people per thousand during each of the six years before the last census in 1976. Heavy out-migration, coupled with high infant mortality rates (around 15 per cent during the first five years of life), resulted in a population growth rate of just under 1.0 per cent during the 26 years between the two most recent censuses. This compares with 2.05 per cent for the country as a whole, and is again the lowest of the Departments, such that Potosi's share of the national population fell from one-fifth to one-seventh during this period. In 1976, only 15.3 per cent of the population of 658,000 was living in towns of more than 20,000 inhabitants, and a further 13.7 per cent in settlements with 2000 to

20,000 inhabitants, making Potosi one of the least urbanized
Departments of the country.

THE ECONOMIC BASE

Potosi's original wealth came from the rich deposits of silver
discovered by the Spanish in the sixteenth century. Today these
deposits are largely exhausted, but mining still represents the principal economic activity of the region, principally tin, accounting in 1976
for some 13 per cent of the work-force and almost a third of regional
product. Agriculture, however, still occupies some 57 per cent of the
labor force, most of whom are farmers in the more temperate valley
areas to the east of the Department, though many raise livestock,
chiefly sheep, goats and llamas, in the high plateaus among the
mountains to the west.

Since mining has long been a major source of export earnings for
the country, the government has invested large sums in the development of this activity, much of it in the Potosi region. On the other
hand, since the agricultural potential of the region is much inferior to
other parts of the country, this sector has been largely ignored by
national planners, with the exception of livestock, specifically llamas.

The varying emphasis placed on these two sectors in the Potosi
region explains the differential patterns of growth. Broadly speaking,
it is the mining areas of the region which have seen what modest
growth there has been, while the agricultural areas for the most part
have lagged or even declined. As Table 1 indicates, 15 provinces in
the Department had population growth rates above the regional average, and four of these were mining areas, and the ten with below
average growth rates, eight had little or no mining activity.

TOWNS AND HINTERLANDS

Of interest here, however, are the interactions between mining and
agriculture, and their effect on the growth of towns and their hinterlands. Table 2 breaks down total population growth in the fifteen
provinces in the Potosi region into rural and urban components
(urban here being places with more than 500 inhabitants), and classifies provinces according to four categories according to the rates of
growth of their urban and rural populations. This shows that most

Table 1.   Population Growth 1950-76 and Mining Activity by Provinces

Employment Location Quotient Mining 1976	Population Growth 1950-76		
	Greater than: Average (0.99 per cent per annum)	Less than:	
Greater than 1.0	Frias Bustillos S. Chichas S. Lipez	N. Chichas Quijarro	
Less than 1.0	Chayanta	Saavedra Charcas Ibanez N. Lipez	Linares Omiste Campos Bilbao

	Degrees of Freedom	Value	Probability
Chi Square Statistic	1	5.000	0.025

provinces experienced either faster urban growth and slower rural growth, or vice versa. Only two provinces had slow growth in both sectors, and only two others had rapid growth in both. In the sections that follow, we seek clues that account for these different patterns, and explain why some towns and hinterlands grow faster, while others lag or decline.

RURAL POPULATION GROWTH AND BASIC ACTIVITIES

First, we want to see what effect the two basic activities of mining and agriculture have had on the growth of rural population. In agriculture we might expect, other things being equal, that areas with higher agricultural productivity would experience faster rates of rural population growth, or put another way, less out-migration. As Table 3 shows, this is what happened in five of the 15 provinces in the Potosi region, and conversely, four areas with low agricultural productivity had low or negative growth rates for rural population.

Table 2.   Urban and Rural Population Growth 1950-76 by Provinces

Rural Population Growth 1950-76	Urban Population Growth 1950-76	
	Greater than: Median (1.03 per cent per annum)	Less than or equal to:
Greater than or equal to median (0.53 per cent per annum)	Frias Campos	Chavanta Charcas N. Lipez S. Lipez Linares Bilbao
Less than median (0.53 per cent per annum)	Bustillos Saavedra N. Chichas S. Chichas Omiste	Ibanez Quijarro

	Degrees of Freedom	Value	Probability
Chi Square Statistic	1	5.402	0.020

Six provinces, however, do not match expectations.   The three in the lower left quadrant all lie to the west of the region in the Altiplano, and have small populations scattered across vast areas earning their living through stock-raising, and small-scale mining operations. Clearly agricultural productivity is not a useful explanatory variable in these cases.

The remaining three provinces in the upper right quadrant are harder to explain.   However, a look at migration data from the 1976 census indicates in two cases that there were substantial outward flows to adjacent areas with large cities — from Saavedra to Frias which includes the City of Potosi, and from Ibanez to nearby Oruro. In Bustillos, the absolute decline of rural population was more than offset by a massive increase in urban population, following the expansion of mining operations in the Llallagua-Siglo XX area. Evidently, in these latter three cases, the lure of more attractive jobs

Table 3.   Rural Population Growth 1950-76 and Agricultural
            Productivity by Provinces

Agricultural Productivity	Rural Population Growth 1950-76	
	Greater than or equal to: Median (0.53 per cent per annum)	Less than:
Greater than or equal to median (40.8 Ha/Na)	Frias Chayanta Charcas Bilbao Linares	Bustillos Saavedra Ibanez
Less than median	N. Lipez S. Lipez Campos	N. Chichas S. Chichas Omiste Quijarro

Chi Square Statistic	Degrees of Freedom	Value	Probability
	1	0.045	0.833

in nearby cities outweighed relatively productive farming opportunities in their rural hinterlands.

    While the association between agricultural productivity and
rural population growth is quite direct, the links between mining and
rural population growth are more complex. *A priori,* we might expect
to find two distinct impacts: (1) a positive one (the spread effect)
through increased demand from urban centers for agricultural
produce, generating additional income to support more labor; and (2)
a negative one (the backwash effect) through increased out-migration
as rural workers are lured to better paying jobs in the growing
mining centers. The net impact in each case would clearly depend on
the interplay between these two factors.

    Table 4 compares the growth of rural population in each
province with the level of mining activity as represented by
employment location quotients. As may be seen, in four of the six

Table 4.   Rural Population Growth 1950-76 and Mining Activity by Provinces

Employment Location Quotient for Mining 1976	Rural Population Growth 1950-76	
	Greater than or equal to: Median (0.53 per cent per annum)	Less than:
Greater than 1.0	Frias S. Lipez	Bustillos N. Chichas S. Chichas Quijarro
Less than 1.0	N. Lipez Campos Chayanta Charcas Bilbao Linares	Omiste Saavedra Ibanez

Chi Square Statistic	Degrees of Freedom	Value	Probability
	1	0.417	0.519

provinces in which mining is important, the rural population grew slowly, while in Frias and S. Lipez it grew faster. In the nine provinces with little or no mining activity six experienced relatively strong rural population growth. The three remaining cases had weak or negative rural growth rates. As mentioned above, Ibanez and Saavedra lost population due mainly to out-migration to adjacent mining centers. By contrast in Omiste out-migration was partly to the thriving trading town of Villazon on the border with Argentina and partly to the neighboring regional center of Tarija.

The evidence suggests, therefore, that the backwash effects of growth in mining activity are generally stronger than the spread effects. Although growing mining centers clearly create demand for agricultural produce, the benefits do not necessarily accrue to the immediate hinterland, or even to adjacent areas. Much depends on the production potential of the surrounding areas, and the relative cost of supplies from producers elsewhere.

Table 5.   Non-Basic Employment 1976 and Mining Activity by Provinces

Employment Location Quotient for Mining 1976	Employment Location Quotient for Non-Basic Activities 1976	
	Greater than 1.0	Less than 1.0
Greater than 1.0	Frias Bustillos S. Chichas Quijarro	N. Chichas S. Lipez
Less than 1.0	Omiste	Saavedra Linares Chayanta Charcas Ibanez Bilbao N. Lipez Campos

	Degrees of Freedom	Value	Probability
Chi Square Statistic	1	5.000	0.025

NON-FARM EMPLOYMENT AND BASIC ACTIVITIES

While the association between mining and rural growth is obscured by the interplay of opposite effects, the link between mining and non-basic employment (defined here as any job other than in the basic activities of agriculture and mining) is much more straight-forward.

Since we have no data on the growth of non-basic employment, the analysis that follows is based on data from the most recent Census in 1976. Table 5 compares provinces in terms of their level of specialization in mining and non-basic jobs using employment location quotients as the indicator. As may be seen, areas in which mining is important also tend to have relatively more non-basic jobs. In part, this is because mining spurs the growth of urban centers by

Table 6.   Non-Basic Employment 1976 and Per Capita Income by Provinces

Per Capita Income 1976	Employment Location Quotient for Non-Basic Activities 1976	
	Greater than 1.0	Less than 1.0
Greater than the median ($389/cap)	Frias Bustillos S. Chichas Quijarro Omiste	S. Lipez Campos
Less than or equal to the median ($389/cap)		Saavedra Linares Chayanta Charcas Ibanez Bilbao N. Chichas N. Lipez

Chi Square Statistic	Degrees of Freedom 1	Value 11.250	Probability 0.0001

concentrating population there, and thus attracting associated urban services.   Conversely, those provinces with little or no mining also have fewer non-basic activities.   The one exception here, Omiste, owes its position to the border trading center Villazon.

NON-FARM EMPLOYMENT AND INCOME LEVELS

A central thesis underlying the scenario outlined at the outset of this paper is that rising rural incomes aid the growth of non-basic jobs in urban areas.   We have no measure of rural incomes *per se*, of their change over time, but we can look at the links between combined rural and urban income levels as a whole for each province and the relative size of the non-basic sector.

As Table 6 shows, the link is strong; provinces with higher income levels also have more non-basic employment.   Only two cases deviate from expectations, Campos and S. Lipez, both of them sparsely populated areas with too small a population to support many urban services.   There is, however, an element of auto-correlation here.   Rising income levels clearly raise demand for non-farm goods and services; but provinces with a large proportion of non-basic jobs are also more likely to have higher average income levels, since earnings are usually higher than in agriculture.

MINING TOWNS AND FARM TOWNS

The strong link between the underlying base of the local economy and the growth of urban population is seen most clearly when we examine individual towns.   Table 7 classifies urban settlements as regional service centers, farm towns, or mining towns, and compares their population growth rates as a group.

The differences between the farm and mining towns stand out sharply.   While the population of farming towns grew by only 0.64 per cent per annum over the 26-year period between censuses, mining towns as a group grew by a much more dramatic 6.51 per cent per year.   Half the farming towns listed in 1950 actually lost population during this period, two of them disappearing from the records entirely.   On the other hand, only two mining towns decline, including the sizable shrinkage of Pulacayo as mineral deposits were exhausted.   The vast majority of them — 26 of the 31 listed in 1976 — came into existence as urban settlements after 1950, as new mines were opened up throughout the region following the nationalization of mining activities after the revolution in 1952.   As a result, the share of the Department's urban population living in farm towns fell sharply from 31 per cent in 1950 to only 14 per cent in 1976, while mining towns doubled their share from 23 to 46 per cent.

While mining towns generally grew faster than others, the interesting question to ask is why some farm towns did better than others.   An examination of the available evidence points to two key factors: productive potential, and effective demand.   In the first

Table 7. Population Growth in Urban Settlements: Regional Service Centers, Farm
         Towns, and Mining Towns 1950-76

Town Type	PU50[1]	%Pu50[2]	PU76[3]	%Pu76[4]	dP50-76[5]
Regional service centres	32,980	46.0	72,974	39.8	3.10
Farm towns	22,351	31.2	26,412	14.4	0.64
Mining towns	16,288	22.7	83,940	45.8	6.51
Total	71,619	100.0	183,326	100.0	3.68

1. PU50   = 1950 population of all towns in the group.
2. %PU50   = percentage share of total 1950 urban population.
3. PU76   = 1976 population of all towns in the group.
4. %PU76   = percentage share of total 1976 urban population.
5. dPU50-76   = average annual percentage increase of population 1950-76.

place, farm towns are clearly not going to grow unless their rural hinterlands have the potential to yield surplus produce of some kind for market. But this alone is not a sufficient condition, as has been plainly demonstrated by a number of towns in the northeast of the region, which all lost population even though farming conditions there are relatively favorable.

Equally important, if not more so, is effective demand from urban centers for the area's produce, which in turn is related to physical accessibility to the producing area. Table 8 compares the population growth of 29 farm towns with a measure of accessibility, based on travel times from the town's hinterland to surrounding urban centers (Dickey and Evans, 1980). As this shows, seven of the nine more inaccessible towns lost population. Of the two exceptions, Kilpani was the site of a new hydro-electric scheme, and Colcha K owes its growth more to a number of plants processing minerals from the Uyuni salt lake.

Of the 20 more accessible towns, 13 of them gained population. All but one of these are located in relatively productive agricultural areas, and within an hour's drive or so of a larger regional center, such as the City of Potosi, Llallagua/Uncia, or Sucre. Seven of the more accessible towns lost population, and although six of these lie on main roads, they are somewhat further away from the regional centers. Three of them are within a few kilometers of Colquechaca, which seems to have grown at their expense. Three others — Cotagaita, Vitichi, and Sacaca — are market towns on better roads with good access to larger regional centers, but they suffer from weak links to their hinterlands.

URBAN FUNCTIONS IN MINING TOWNS AND FARM TOWNS

Given that mining towns generally have been growing faster than farm towns, how has this affected the distribution of public facilities and the diffusion of commercial services among settlements? Are mining towns also attracting a wider range of urban services and activities, or not? Table 9 shows how certain kinds of urban functions are distributed among different types and sizes of towns. Functions are clustered under four groups: mining activities, agricultural services, commercial activities, and public facilities.

The first thing to notice, as might be expected, is that the range of urban functions increases with population size, rising from an average of 13 for villages with less than 500 inhabitants, to 40 for the eight towns with more than 5000 people. There are no farm towns in the largest category, but interestingly, in each of the other three categories, farming towns tend to possess a wider range of urban functions than mining towns of a similar size. This is especially noticeable in the 1000 to 5000 population size group, where farm towns have an average of 22 function types versus 17 for mining towns.

In general public facilities, such as schools, health services, and infrastructure, appear to be fairly evenly distributed across different sizes and types of towns. The largest towns possess an average of 15 functions, while the smallest villages have seven, and farm towns are only slightly favored. The range of commercial services, on the other hand, varies more widely, rising from only four functions among the smallest villages to 20 in the largest towns, and farm towns tend to have a greater range than mining centers.

This indicates that the distribution of urban functions is deter-mined not so much by the population size of the settlement itself, or

Table 8.   Urban Population Growth 1950-76 and Accessibility by
            Farm Towns

Total Per Capita Accessibility	Urban Population Growth 1950-76	
	Positive	Negative
Levels 3 and 4 (>20.0 per cent maximum)	Catavi Civil	Aymaya
	Chaqui	Sacaca
	Chiracoro	Macha
	Chayanta	Pocoata
	Betanzos	Cotagaita
	Mojocorillo	Ocuri
	Colquechaca	Vitichi
	Puna	
	Caripuyo	
	Ravelo	
	Caiza D	
	Chuafaya	
	Colavi	
Levels 1 and 2 (<20.0 per cent maximum)	Colcha K	Tinguipaya
	Kilpani	Panacachi
		Acacio
		Arampampa
		S.P. de Buena Vista
		Torotoro
		Chayrapata

	Degrees of Freedom	Value	Probability
Chi Square Statistic	1	2.885	8.089

its growth rate, but the market area of the town together with its
rural hinterland.   Despite the slower growth rate of farm towns, they
still possess a larger range of functions, since they tend to serve
larger hinterland populations than mining towns of a similar size.

Table 9. Urban Functions in Mining Towns and Farm Towns

Town Size	Number of Towns	Average Population	Urban Functions				
			Mining	Agriculture	Commerce	Public Service	Total
More than 5000							
Service towns	5	23,517	1.8	3.8	23.0	16.6	45.2
Farm towns	0	0	0	0	0	0	0
Mining towns	3	13,573	1.7	2.7	15.0	11.7	31.0
Subtotal	8	19,788	1.8	3.4	20.0	14.8	39.9
1000 - 5000							
Farm towns	7	1843	0.7	1.6	9.3	10.4	22.0
Mining towns	14	2301	1.5	0.9	6.4	8.6	17.3
Subtotal	21	2149	1.2	1.1	7.3	9.2	18.9
500 - 1000							
Farm towns	9	756	0.4	0.9	6.4	8.8	16.6
Mining towns	5	792	1.6	0.6	4.8	7.8	14.8
Subtotal	14	769	0.9	0.8	5.9	8.4	15.9
Less than 500							
Farm towns	11	331	0.6	0.4	4.3	7.6	12.9
Mining towns	3	308	1.7	0.0	4.0	6.3	12.0
Subtotal	14	326	0.9	0.3	4.2	7.4	12.7
All towns - Total	57	3838	1.1	1.1	8.0	9.3	19.6

DISTRIBUTION OF AGRICULTURAL AND COMMERCIAL SERVICES

Can we discern any pattern to the distribution of certain kinds of urban functions? *A priori*, we might expect to find, for example, that agricultural services would tend to concentrate in areas which specialized in farming. Table 10 compares the distribution of agricultural services with a measure of agricultural productivity in each province.

As may be seen, there is little if any correlation between the two. Only three of the seven more productive farming areas are relatively well provided with agricultural services, and these all lie on main roads within easy reach of larger cities. Four of the better farming ares are poorly provided with agricultural services, and three of these lie in the remote north-eastern corner of the department. On the other hand, half the less productive farming areas possess a relatively wide range of agricultural services, two of them, Sud Chichas and Omiste, include within their boundaries the regional centers of Tupiza and Villazon, which clearly serve extended market areas beyond the provincial boundary. The key factor explaining the distribution of agricultural services, then, is not so much the level of agricultural activity or productivity, as the level of accessibility of the towns to their potential markets.

In the case of commercial activities, we might expect to find that higher income levels generate demand for a greater diversity of such services. Table 11 compares the distribution of commercial services among towns in each province with per capita income levels.

In this case, the link is a little stronger than for agricultural services. The link is clearest among the provinces in the lower right quadrant, and the upper left quadrant, which includes the five major regional centers of Potosi, Llallagua/Uncia, Tupiza, Uyuni, and Villazon. Four cases run counter to expectations. Campos and Sud Lipez are areas with extremely small populations scattered over large areas. Ibanez is well linked to the regional center of Oruro. But the real surprise is Charcas, which is so hard to reach. The main town here is San Pedro de Buena Vista, which lost half its population between 1950 and 1976. It seems though, that despite this hemorrhage, the town still continues to serve a large and somewhat captive hinterland population.

Table 10.   Distribution of Agricultural Services[1] and Agricultural
            Productivity by Provinces

Agricultural Productivity	Distribution of Agricultural Services (average number of services per town)	
	1.0 or more	Less than 1.0
Greater than median (40.8 Ha/Na)	Frias Saavedra Ibanez	Bustillos Chayanta Charcas Bilbao
Less than or equal to median	N. Chichas S. Chichas Linares Omiste	N. Lipez S. Lipez Campos Quijarro

	Degrees of Freedom	Value	Probability
Chi Square Statistic	1	0.134	0.714

1.   Agricultural services include: weekly market, daily market, slaughterhouse, farm
supplies store, farm tools and equipment store farm machinery hire or sale, silo or storage
facility.

CONCLUSIONS

The purpose of this rudimentary analysis of the Potosi region was to
determine which factors explain why some small towns in a region
grow while others do not.   Such information is needed by planners
and policy makers in formulating policies and strategies for
promoting small towns, regional development, and urban
deconcentration.   The main conclusions are as follows.

Small towns have little scope for growth unless they or their
hinterlands can count on a real or potential resource base of some
kind.   This point has been well documented in numerous earlier
studies (notably Perloff and Wingo, 1961), but is illustrated once
again in the case of Potosi.   This resource constitutes a comparative
advantage of some kind, and this may vary from one place to another

Table 11.   Distribution of Commercial Services and Per Capita
             Income by Provinces

| Per Capita Income 1976 | Distribution of Commercial Services (average number of service types per town) | |
	More than 6.5	Less than 6.5
Greater than the median ($389/cap)	Frias Bustillos S. Chichas Quijarro Omiste	S. Lipez Campos
Less than or equal to the median ($389/cap)	Charcas Ibanez	Saavedra Linares Chayanta N. Lipez Bilbao N. Chichas

	Degrees of Freedom	Value	Probability
Chi Square Statistic	1	5.402	0.020

and from one time to another. Silver was the original resource of the
Potosi region, but latterly it has bee replaced by tin and other
minerals.   Thus, within the department, it is those towns based on the
exploitation of mineral resources that have grown the fastest, since
this is the region's strongest suit in world markets.   Although some
areas of Potosi possess limited agricultural potential, this is difficult
to exploit due partly to more productive possibilities elsewhere in the
country.

A resource base, however, is a necessary but not sufficient condition for growth to take place: there has to be effective demand, or
more properly effective supply to take advantage of that demand.
The mining towns grew not merely because they possessed a

resource, but also because the national government has consistently given high priority to the exploitation of mineral resources, which have long been a major source of foreign exchange earnings for the country.

Thus, a second factor which features prominently in determining the pattern of growth among small towns in a region is the nature of national development priorities and sectoral policies. More than anything else, the dramatic variations in the growth of smaller towns in the Potosi region is explained by national priorities that accorded first place to the mining sector, while largely ignoring agriculture. Of total public and private investment in the region over the past 30 years, by far the greatest part has been for mining and related activities (Cupe, 1981). Not surprisingly, therefore, we find that mining towns have captured the lion's share of growth in the region, while farming towns have languished or declined.

Effective supply also depends crucially on physical accessibility. If unit production costs for farm goods do not vary much across the region, then the critical factor becomes unit transport costs. Areas that are more distant from the main market are handicapped, especially as in many parts of Potosi where poor road conditions make for long trip times and high transport costs. These high transport costs may be reduced in two ways: by improving roads, thus reducing travel times; and by bulking produce thus reducing unit costs.

The evidence from Potosi demonstrates clearly that if farm towns are to grow, they must be competitive in supplying produce to large urban markets. Villages close to urban markets can compete by shipping only small amounts of produce, but more distant towns need to ship larger volumes to remain competitive. This means that if more distant farming towns are to compete successfully, they must not only be accessible to larger urban markets, but they must also be adequately linked with producers in their hinterlands, in order to function efficiently as centers for bulking marketable farm produce.

Acknowledgement

The author acknowledges the valuable contribution made in preparing data and earlier analyses on which this paper is based by a team headed by Sr. Alfredo Bellott, previously Chief of the Planning

Department of CORDEPO, the regional development corporation for the
Department of Potosi, Bolivia.

BIBLIOGRAPHY

Cordepo, 1981, Funciones urbanas en el desarrollo Rural: Resultados
    del Estudio en Potosi, Potosi: CORDEPO, Vols. I and II.
Cupe, German, 1981, Plan regional de desarrollo 1982-86: Analisis
    Global, Potosi: CORDEPO.
Dickey, John and Evans, Hugh, 1980, "A Technique to Help Evaluate
    Function/Linkage Packages," US Agency for International
    Development, Office of Urban Development, Washington, DC.
Evans, Hugh, 1982, "Urban Functions in Rural Development: The Case
    of the Potosi Region in Bolivia," US Agency for International
    Development, Regional and Rural Development Division, Vols. I
    and II. Washington, DC.
Gibb, Arthur, Jr., 1984, Tertiary urbanization: The agricultural market
    centre as a consumption-related phenomenon, Regional
    Development Dialogue, Vol. 5, No. 1 (Spring 1984):110-43.
Government of Kenya, 1986, Rural-urban balance, chapter 4 in:
    "Economic Management for Renewed Growth," The Government
    Printer, Nairobi.
Hansen, Niles, 1982, The role of small and intermediate cities in
    national development processes and strategies, chapter 14 in:
    "Small Cities and National Development," O.P. Mathur, ed., United
    Nations Centre for Regional Development, Nagoya, Japan.
Jones, Barclay Gibbs, 1986, Urban support for rural development in
    Kenya, Economic Geography, Vol. 62, No.3. (July 1986): 201-14.
Ligale, Andrew Ndooli, 1982, "The role of small and intermediate size
    cities in national development in Africa," unpublished paper.
Mathur, Om Prakash, 1982, "Small Cities and National Development,"
    United Nations Centre for Regional Development, Nagoya, Japan.
Perloff, Harvey, and Wingo, Lowden, 1961, Natural resource endow-
    ment and regional economic growth, chapter 12 in: "Regional
    Policy," John Friedmann and William Alonso, eds., MIT Press,
    Cambridge, MA.
Rondinelli, Dennis A., 1981, "Developing and Managing Middle-Sized
    Cities in Less Developed Countries," USAID, Office of Urban
    Development, Washington, DC.
Rondinelli, Dennis A., 1983, Towns and small cities in developing
    countries, Geographical Review, Vol. 73, No.4 (October 1983).

Rondinelli, Dennis A., and Evans Hugh, 1983, "Integrated regional development planning: Linking urban centres and rural areas in Bolivia," *World Development,* Vol. 11, No. 1: 31-53.

United Nations Centre for Human Settlements, 1985, "Planning and Management of Human Settlements with Emphasis on Small and Intermediate Towns and Local Growth Points".

10
Urban Centrality and Agricultural Productivity: Regional Development in the Hashemite Kingdom of Jordan

Peter L. Doan
Program in Urban and Regional Studies, Cornell University

INTRODUCTION

Regional development is a concern of many policymakers in developing countries. The government of the Hashemite Kingdom of Jordan has recently renewed its emphasis on regional planning in conjunction with its preparations for the latest Five Year Plan (1986-90). The Committee overseeing this work was chaired by Crown Prince Hassan who has repeatedly expressed his desire to eliminate "pockets of poverty" by stimulating development in the least developed regions of the country.

Such efforts are equally important for those concerned with the practical elements of regional development policy and for those concerned with developments in urban and regional theory. Of particular interest to both groups is the role which secondary cities can play in stimulating wider regional development. Previous work has indicated substantial potential for development strategies based on the medium-sized cities in an urban system (Mathur, 1982; Rondinelli, 1983), however the policy mechanisms for such development have not yet been fully explored. The role of this paper is to examine the case of Jordan and analyze the importance of linkages between agriculture and urban economic activity based in intermediate towns.

The argument of this paper is that there is considerable potential for stimulating regional development through policies designed to strengthen rural-urban linkages in less-developed areas. Linkages between market towns and their agricultural hinterlands make intermediate inputs accessible to agricultural producers and result in higher productivity. These increases in agricultural productivity in turn will stimulate demand on the part of agricultural producers for consumption goods.

Attention is given first to the existing development context in Jordan in which over the last 50 years both the rates of urbanization and increased agricultural productivity have been extremely high. Then, after elaborating a brief model of rural-urban relationships the focus is on existing cross-sectional data available for 1985. The results of several rank order correlations are used to analyze the existing relationships between urban centrality and agricultural productivity in Jordan.

The cross-sectional data were collected as part of a national village survey conducted by the Regional Planning Department of the Ministry of Planning to gather information on the location of population and economic activities in all villages in preparation for the Five Year Plan. The author worked as an advisor to this department under the Settlements and Resource Systems Analysis Project (SARSA) of Clark University which is funded by USAID. An informant survey was conducted in each village and town in the country in which the mayor (*mukhtar*) of each village was intensively interviewed. This information was supplemented by data from central government ministries about their activities in each locality. Data on business activities were obtained from the Chambers of Commerce of all towns with a population in excess of 5000 and from the *mukhtars* of all other places.

DEVELOPMENT CONTEXT

Jordan is located at the junction of two trade routes at the western edge of an arid plateau stretching south to Saudi Arabia and east to Iraq (Figure 1). This location has created considerable tension between settled agriculturalists in the areas with higher rainfall and the nomadic bedouin tribes that move through the region with their livestock herds. The productivity of agriculture is linked to the ability of social and economic institutions to ensure the security of regular harvests and to protect investments in agricultural infrastructure (Jones, 1985). The history of the region is marked by

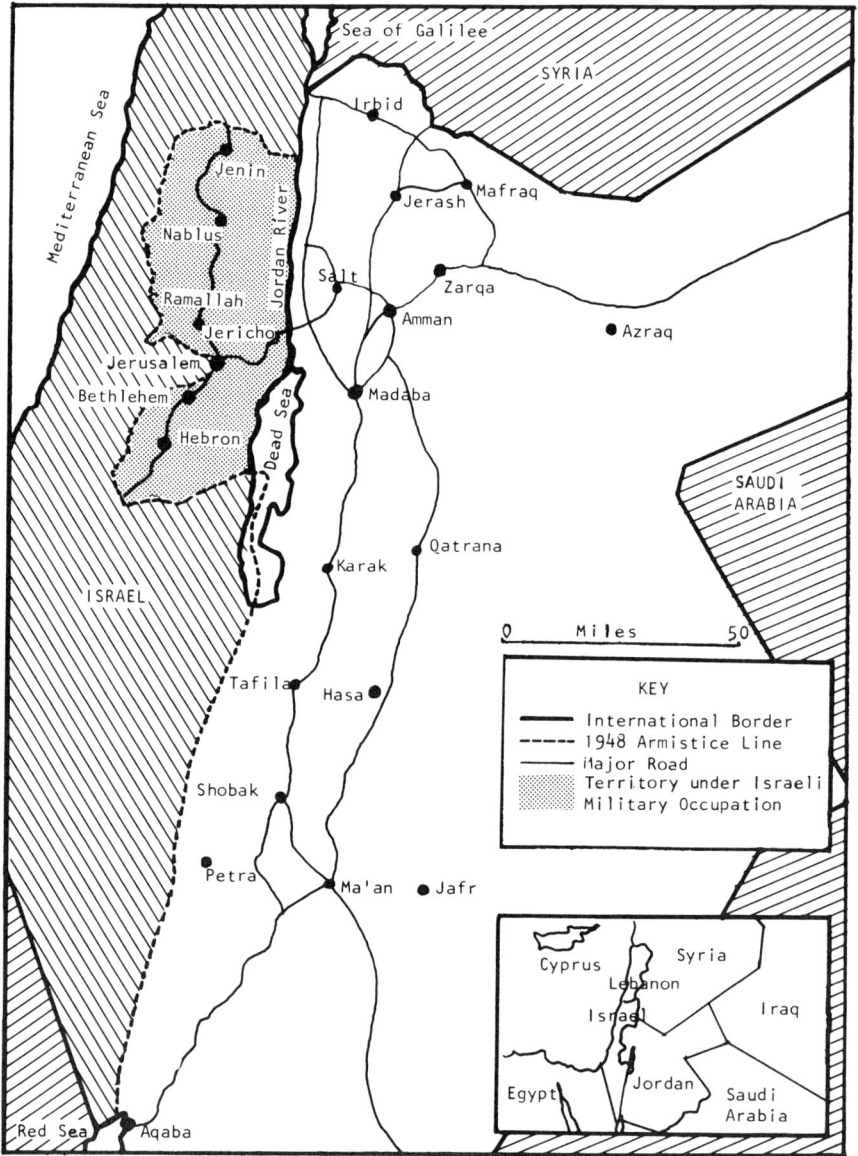

Figure 1. Map of Jordan. (Figure reproduced from Papers and Proceedings of Applied Geography Conferences, Vol. 9 (1986) with the kind permission of the author.)

Table 1.   Population Growth in the Hashemite Kingdom of Jordan, East Bank: 1943-85

Area	Year				
	1943 (Est)	1952 (Census)	1961 (Census)	1979 (Census)	1985 (Census)
Irbid	180,000	178,504	273,976	611,280	632,345
Mafraq[1]					93,448
Amman	100,000	282,516	433,618	1,173,170	1,095,543
Zarqa[2]					424,924
Balqa[3]		51,508	79,057	147,827	180,099
Kerak	40,000	43,790	67,211	125,959	112,541
Tafila[4]					37,371
Ma'an	20,000	30,566	46,914	74,761	92,901
Total	340,000	586,885	900,776	2,132,997	2,669,172

1.   Mafraq is with Irbid until 1985.
2.   Zarqa is with Amman until 1985.
3.   Balqa is with Amman for 1943 only.
4.   Tafila is with Kerak until 1985.
Sources: Konikoff, 1946; Census, 1952, 1961, 1979; National Village Survey.

periods when security was provided and agricultural and commercial activities flourished.   At other times both economic activities and the population of the region itself were greatly diminished.   The relative neglect of the region by the Ottoman Empire in the mid-nineteenth century is generally considered the nadir of the economic and population growth curves.

With independence in 1922 the Emirate of Transjordan began to reestablish the conditions necessary for economic activity.   However the existing rural-urban balance was drastically altered by the turmoil resulting from the creation of the state of Israel in 1948.   The Arab-Israeli wars of 1948 and 1967 displaced large numbers of Palestinians, many of whom fled as refugees to the East Bank.   Table 1 displays the growth of population by region from 1943-85.   During this period the population of the East Bank of Jordan increased from 340,000 to 2.6 million.   These massive population flows created severe distortions in the urban system.   The city of Amman, for example, was elevated from a sleepy town of 30,000 in 1943 (Mazur, 1979) to a bustling capital of 108,304 people in 1952 (Census of Housing quoted by Hacker, 1960) to its current population of 812,513 (Ministry of Planning estimates based on Census of 1979).   In 1952 the percentage of urban population was 29 per cent and increased substantially by 1961 to 58.8 per cent.   By 1979 the percentage had reached the extremely high level of 70.2 per cent, which is comparable to levels of urbanization in the United States.

Table 2 demonstrates the effects of this growth on the distribution of population by size of locality based on Ministry of Planning estimates for 1985.   This distribution reveals that 73.9 per cent of the population of Jordan live in towns larger than 5000 inhabitants.   Amman itself accounts for 30.4 per cent of the total population and the greater Amman area accounts for more than 50 per cent.

During this same period the agricultural production system was undergoing rapid expansion and modernization.   The intensity of agriculture production increased dramatically with the rapid population influx to the East Bank.   However at the same time the efficiency of agricultural production has increased, especially in the irrigated areas, so that the agricultural labor force declined as a percentage of the economically active population.   In 1955 a World Bank study team estimated that there were 195,000 workers (63.1 per cent) engaged in agricultural production, with 59,000 in the public

Table. 2.   Distribution of Population by Size of Locality, 1985

Size of Locality	Number of Localities	Percentage of Total	Cumulative Percentage of Total	Population of Size Class	Percentage of Total	Cumulative Percentage of Total
> 500,000	1	0.09	0.09	812,513	30.44	30.44
100,000 - 499,999	2	0.18	0.28	418,014	15.66	46.10
50,000 - 99,999	2	0.18	0.46	117,137	4.39	50.49
20,000 - 49,999	8	0.74	1.20	260,873	9.77	60.26
10,000 - 19,999	10	0.92	2.13	134,386	5.03	65.30
5000 - 9999	34	3.14	5.27	229,153	8.59	73.88
2000 - 4999	98	9.06	14.33	300,647	11.26	85.15
1000 - 1999	123	11.37	25.69	172,510	6.46	91.61
500 - 999	153	14.14	39.83	106,416	3.99	95.60
0 - 499	651	60.17	100.00	117,523	4.40	100.00
Total	1082	100.00		2,669,172	100.00	

Source: National Village Survey Projections Based on 1979 Census.

sector and 55,000 engaged in urban enterprises for a total labor force
of 309,000, excluding unemployed refugees (IBRD, 1957).

As is indicated in Table 3 the percentage of the labor force
engaged in agricultural production declined to 33.5 per cent in 1961
and 9.9 per cent in 1982. At the same time agricultural production,
particularly in the categories of fruits and vegetables, increased
dramatically. In general these are crops which require more extensive investments in infrastructure and an assured marketing system
due to problems of spoilage.

Table 4 presents more information on the acreages and volume
of production for the major crops in Jordan from 1943-82. These
statistics show large fluctuations in area planted and actual yield
which are partially related to political turmoil discussed above and
partially related to variations in annual rainfall. However several
key trends are still evident. In the rainfed regions the area devoted
to production of wheat, lentils, and chick peas has been significantly
reduced and has been replaced by barley, a more drought-resistant
crop, and tobacco, a more lucrative cash crop. In vegetable
production it is evident that for high-value crops like tomatoes and
cucumbers the volume of production has increased at a much faster
rate than the area devoted to those crops, thus indicating
considerable productivity increases.

However, Table 5 shows that even with this increase in productivity the agricultural sector accounted for a declining share of gross
domestic product. This decline indicates that other sectors, particularly manufacturing and construction, have expanded more rapidly
than agriculture. Part of the boom in construction is related to the
increased remittances from the estimated 350,000 expatriate
workers in Saudi Arabia and the Gulf which caused a huge increase in
demand for new housing (Economist Intelligence Unit, *Country Profile
for Jordan 1986-87*, p.7). In addition, it is important to note that as
the agricultural sector becomes increasingly modernized, it uses
larger and larger quantities of purchased inputs from other sectors
which are not accounted for in this table.

The important question arising from this analysis is to what
extent has the increase in population and levels of urbanization contributed to increasing agricultural productivity in outlying regions.
For regional development policies to be developed it is extremely
important to understand the consequences of such rapid urbanization

Table 3.   Distribution of Labor Force by Economic Activity,
            1961 and 1962

Sector	Employment in Thousands (per cent)			
	1961		1982	
Agriculture[1]	72.9	(33.5)	45.0	(9.9)
Mining and quarrying	4.7	(2.2)	6.1	(1.3)
Manufacturing	17.5	(8.0)	39.2	(8.6)
Electricity and water supply	22.2	(10.2)	5.1	(1.1)
Construction[1]	0.9	(0.4)	6.0	(1.3)
Trade[1]	17.4	(8.0)	44.0	(9.7)
Transportation and storage[1]	7.6	(3.5)	32.0	(7.0)
Services	28.7	(13.2)	223.1	(49.1)
Other (not specified)	45.7	(20.1)	54.0	(11.8)
Total	217.8		454.5	

1. Estimated for 1982.
Source: Department of Statistics Annual Statistical Yearbook, 1983.

with respect to agricultural productivity and regional development in
general.   Has the development of the urban system facilitated
regional growth or has the increasing primacy of the system been a
constraint to broader development goals?

ANALYSIS OF AGRICULTURAL PRODUCTIVITY AND URBAN CENTRALITY

Establishing regional policies to strengthen rural-urban linkages and
stimulate regional development is contingent on a more complete
understanding of the way in which these linkages affect agricultural
production and hence have an impact on regional development.
There is considerable scope for raising productivity in agriculture by
making available locally the goods and services which are necessary
for intensified agricultural development.   Strengthening the urban
distribution system and improving the linkages between agricultural

Table. 4.   Area (in Acres) and Tons of Production of Major Agricultural Crops by Year

Crop	1943	1953	1967	1973	1982
			Area (in Acres)		
Wheat	1,400,000	2,216,337	2,300,000	2,441,800	1,036,750
Barley	500,000	762,273	583,900	531,700	657,900
Tobacco	2599	14,538	-	36,400	69,130
Lentils	-	117,903	227,900	241,900	64,043
Chick Peas	-	63,141	36,700	73,400	28,040
Tomatoes	15,300	-	165,300	132,800	180,465
Cabbage/Cauliflower	1800	-	13,700	9300	30,878
Eggplant	7350	-	28,700	20,000	39,911
Melon	21,000	-	68,300	115,700	67,551
Cucumbers	-	-	36,900	20,500	51,815
Olive	-	-	84,600	324,200	304,289
Grape	-	-	50,800	37,300	125,162
Citrus	-	-	14,400	18,100	64,204
Banana	-	-	12,900	5100	10,845
Total area	1,948,049	3,174,192	3,624,100	4,008,200	2,730,983

Tons of Production

Wheat	100,000	94,032	196,100	50,400	62,289
Barley	55,000	29,847	63,400	5900	39,248
Tobacco	97	273	1800	1100	5244
Lentils	10,100	5250	24,200	4800	4284
Chick Peas	2300	1775	3100	2000	2448
Tomatoes	-	-	216,300	83,100	575,424
Cabbage/Cauliflower	-	-	33,300	10,400	11,417
Eggplant	-	-	58,500	14,700	107,922
Melon	-	-	58,300	56,000	45,852
Cucumbers	-	-	28,200	10,400	127,765
Olive	15,800	29,000	22,200	5200	34,263
Grape	25,700	16,395	28,100	22,200	36,017
Citrus	-	1100	29,000	15,400	230,513
Banana	-	5200	22,200	2300	38,870
Total production	208,997	182,872	784,700	283,900	1,321,556

Sources: Konikoff, 1946; Census of Agriculture, 1953; Central Bank Statistical Series 1963-83; National Village Survey 1985.

Table 5.   Industrial Origin of Gross Domestic Product by Year
            (in millions of Jordan dinars)

Industry	1964	1974	1983
Agriculture, forestry and fishing	34.1	30.3	99.1
Mining and manufacturing	12.3	40.5	256.8
Electricity and water supply	1.0	3.0	28.5
Construction	5.5	16.8	126.8
Wholesale/retail trade, restaurants and hotels	28.0	42.3	233.7
Transport and communications	12.0	22.8	146.3
Financing and real estate services	1.5	22.5	144.8
Producers of government services	19.7	54.3	232.9
Other services	21.4	9.9	49.1
Total	135.5	242.4	1318.0

Source: Central Bank of Jordan Yearly Statistical Series (1964-83).

areas and the market towns which serve them is an important means of achieving this end (Jones, 1986).

A wide variety of intermediate inputs are necessary to sustain agricultural production at high levels of productivity. When more goods and services are readily available, farmers will be able to take advantage of these inputs and increase their productivity, resulting in higher levels of production and greater value of agricultural product. In addition the outputs of agriculture will be more readily available for use by other sectors as intermediate inputs in the production of final demand goods. A secondary benefit to regional development would be the resulting increase in demand for consumer goods as a result of increasing agricultural incomes in less developed regions (Gibb, 1984).

In a perfectly rational world this model would predict that variations in the productivity of farmers would be closely related to differences in the availability of goods and services.   The actual relationship between urban centrality and agricultural productivity is complicated by several factors and could follow any of four possible patterns.   Figure 2 illustrates the range of possibilities.   Cells 1 (low productivity — low centrality) and 4 (high productivity — high centrality) are well explained by the model elaborated above.   Cell 3 (low productivity — high centrality) includes those cases in which other reasons for high centrality values exist, such as a break of bulk point for transhipment of goods or a natural resource extraction area with little agricultural potential.

Cell 2 (high productivity — low centrality) presents a more complex case which is possible in the case of plantation agriculture or in situations where government policy has provided sufficient infrastructural support to stimulate production, without providing additional means for wider development.   Under the right climatic conditions such investments may result in high levels of productivity, however the implications for broader regional development are quite dim.

To analyze the interactions between agricultural areas and intermediate towns it was necessary to begin by delineating the spatial boundaries of each area.   The agricultural hinterland of each market town was defined by the SARSA team using spatial inter-action data which linked agricultural areas to the market town which most frequently provided the basic goods and services to the rural population.   Each of these groupings (development clusters) was then joined into a larger subregion using a similar procedure (see Honey et al, 1986 for further detail).   This paper presents the preliminary results of analysis conducted at the subregion level.

Several variables were constructed with the subregion as a unit of analysis.   The concept of centrality drawn from the literature (Berry and Garrison, 1958a; Davies, 1967; Marshall, 1969) is used as a measure of the availability of economic goods and services in urban places.   The imputed farmgate value of agricultural production is used as a measure of agricultural activity in rural areas.   We would expect to find that centrality is positively correlated with higher values of agricultural production.   Similarly areas with higher settle-ment density have correspondingly higher levels of agricultural pro-duction.

| | | Agricultural Productivity | |
		Low	High
	Low	1	2
Centrality			
	High	3	4

Figure 2.   Contingency table of possible outcomes.

Value of Agricultural Production

This variable was computed for each locality using the agricultural data from the National Village Survey. It was then aggregated to the cluster and subregional level. The data included estimates of production totals for field crops, fruits and vegetables, and a variety of livestock (cows, chickens, goats, and sheep). The value used is the imputed farmgate value.

Centrality

A measure of centrality was computed following Davis (1967). This index number is a proxy for the availability of a wide range of economic goods and services in each area. The index is related to the location coefficients used by industrial location analysts. The centrality index is calculated as follows:

$$C_i = \left[\sum_{f=1}^{n} \frac{1}{T_f} \times 100 \right] N_f,$$

where C_i = centrality index for place i,
      T_f = total number of firms of function f in the system,
      N_f = number of firms of function f in place i.

The total centrality index of a place is the summation of the centrality scores for all functions in that place.   The total centrality of a cluster or subregion is likewise the sum of the centrality scores for all places within the area.

Number of Market Towns

The number of towns with population in excess of 1000 is used to measure both the settlement density in a region and the extent to which competing towns may exist within a cluster and within a sub-region.   In a theoretically correct central place system within a given area the number of towns for each size class is clearly specified.   The extent to which a system deviates from the theoretically expected settlement pattern will influence the availability of goods and services throughout the area.   Thus a region with a high aggregate centrality score in a single urban area may have less accessible goods and services than another region with lower aggregate score, but more nodal distribution points.

Results of Rank Order Correlation Calculations

Table 6 presents the results of the Spearman's rank order correlation calculations.   These correlations are discussed below.   The formula used in this calculation is as follows:

$$\text{Spearman's } r = 1 - \frac{6 \, \Sigma \, (D_i)^2}{N(N^2\text{-}1)} \, ,$$

where D_i = difference between i^{th} observation rank for the two variables.

Centrality and Population

In keeping with the findings of other researchers (Berry, Thomas, Stafford, and Marshall) there is a highly significant relationship between the level of centrality (TCI) and the population (POP) of each region.   A correlation of 0.93 was computed between population and centrality at the subregional level.

Table. 6. Rank Order Correlations by Subregion in Jordan

Name	Rank VAP	Rank Towns	Rank TCI	Rank POP	$(D_i)^2$ pop-tci	$(D_i)^2$ vap-town	$(D_i)^2$ vap-tci	$(D_i)^2$ town-tci
Amman	8	3	1	1	0	25	49	4
Madaba	4	12	4	4	0	64	0	64
Naur	16	17	17	21	0	1	1	0
Mowaqqar	13	20	18	19	0	49	25	4
Irbid	11	4	3	3	16	49	64	1
Mazar Sham.	9	1	13	6	49	64	16	144
Ramtha	22	19	11	12	1	9	121	64
Hartha	23	6	24	17	4	289	1	324
Irbid Ghor	1	10	14	11	9	81	169	16
Sammo'	17	8	20	13	4	81	9	144
Ajlun	10	5	12	9	9	25	4	49
Jarash	6	7	9	5	1	1	9	4
Balqa Ghor	2	2	15	14	49	0	169	169
Salt	12	16	6	7	1	16	36	100
Ain el Basha	18	22	10	8	64	16	64	144
Kerak	19	14	7	15	121	25	144	49
Qasr	20	18	23	24	49	4	9	25
Mazar Janoob.	14	13	22	20	4	1	64	81
Ayy	27	23	27	27	1	16	0	16

Region								
Kerak Ghor	21	21	28	25	4	0	49	49
Qetraneh	31	30	32	33	16	1	1	4
Aqaba	30	32	5	16	1	4	625	729
Quaira	26	31	29	30	16	25	9	4
Shobak	25	26	25	28	1	1	0	1
Wadi Moosa	28	27	26	26	9	1	4	1
Ma'an	24	28	19	23	0	16	25	81
Husseiniyyeh	34	33	33	32	0	1	1	0
Wadi Araba	32	34	34	34	9	4	4	0
Mafraq	7	11	8	10	1	16	1	9
Mafraq Desert	5	15	16	18	1	100	121	1
Zarqa	3	9	2	2	0	36	1	49
Tafila	15	24	21	22	1	81	36	9
Bsaira	29	25	30	29	1	16	1	25
Hasa	33	29	31	31	0	16	4	4
$\Sigma (D_i)^2 =$				442	1134	1836	2368	
Spearman's $r =$				0.92830	0.81607	0.70220	0.61592	
Standard normal (Z Score) =				5.33272	4.68796	4.03386	3.53817	
Standard error =				0.17407				

Abbreviations: VAP = Value of agricultural production; TCI = level of centrality; and POP = population of each region.

Value of Agricultural Production and Number of Towns

It was anticipated that there would be a strong relationship between the numbers of market towns and the value of agricultural production. A Spearman's rank order correlation of 0.82 was computed between the value of agricultural production (VAP) and the number of towns with more than 1000 people. This relationship is weaker than that between centrality and population because of Type 3 subregions such as Aqaba which has high levels of centrality as the only port for the entire country, but is bordered on three sides by the desert and therefore has negligible agricultural production.

Value of Agricultural Production and Centrality

To evaluate the role of rural-urban linkages in regional development the most interesting relationship is that between centrality and the value of agricultural production. If the urban distribution system is properly oriented towards the provision of goods and services to agricultural producers, then we would expect a high correlation between these two variables. The Spearman rank order correlation of 0.70 is still significant, but indicates that central goods and services are more directed towards centers of population and are less oriented towards agricultural producers. This is a natural reflection of the concentration of economic goods and services in the major population centers, but as such they are less accessible to agricultural producers especially owner-operated farms in the rural areas. While agricultural production clearly occurs in spite of the uneven distribution of intermediate goods and services, there is substantial potential for economic growth and regional development if the linkages which distribute these functions to agricultural producers can be strengthened.

Number of Towns and Centrality

One would expect that areas with more towns would have more higher order places and therefore greater aggregate centrality. While there is still a significant correlation of 0.62, towns in Jordan do not conform to central place expectations for the delivery of goods and services equally. This is no great surprise since topographic and administrative features are quite important in determining both settlement density and agricultural productivity.

An important example of a Type 2 subregion is the Jordan Valley. This aberration is a result of the creation of the Jordan Valley Authority (JVA) in 1973. The JVA was the mechanism for large scale investments in the East Ghor Canal which vastly increased the area available for irrigated agricultural production and attempted to provide the infrastructure and services necessary to support that activity. However, the Jordan Valley still has lower levels of centrality than one would have expected if urban goods and services were commensurate with the value of agricultural production. The infrastructural investments were sufficient to raise agricultural productivity, but did not provide sufficient urban services for significant urban economic activities to occur within the region.

Furthermore, the JVA as a regional planning entity did not sufficiently emphasize the rural-urban linkages necessary to create the conditions for self-sustaining regional growth. In any case, this high-cost strategy cannot be easily replicated elsewhere in Jordan, both because of financial constraints and because the water resources for such extensive irrigation are quite limited. In order to realize the potential for increased agricultural development in the rainfed zone of Jordan, the intensification of production and the adoption of higher-yielding inputs must be balanced by strengthening the urban distribution system in support of agricultural production.

CONCLUSIONS

This paper has argued that policy makers concerned with regional development should be aware of the importance of rural-urban linkages. These linkages are necessary for the distribution of intermediate inputs to agriculture and the corresponding provision of agricultural outputs as inputs to other producers. Regional policies which do not sufficiently take this into account may result in unbalanced development such as the Jordan Valley. Furthermore, to stimulate development in less productive regions it is important to gear initial investments towards the intermediate-sized market town which is closely linked to agricultural producers; otherwise the distribution system for agricultural inputs and outputs may constrain wider regional development.

More analysis needs to be done on the precise nature of the rural-urban linkages which tie agricultural producers to urban suppliers and producers. A more detailed study of these patterns is currently underway and should reveal more fully the role of the intermediate towns in stimulating agricultural development.

REFERENCES

Aresvik, Oddvar, 1976, "The Agricultural Development of Jordan," Praeger Publishers, New York, NY.
Berry, Brian J. L., and Garrison, William, 1958a, The functional base of central place systems, *Economic Geography*, Vol. 34:145-54.
Berry, Brian J. L., and Garrison, William, 1958b, Recent developments of central place theory, *Regional Science Association, Papers and Proceedings*, Vol. 4:107-20.
Christaller, Walter, 1966, "Central Places in Southern Germany," translated by Carlisle Baskin, Prentice-Hall, Englewood Cliffs, NY.
Davis, Wayne K. D., 1967, Centrality and the central place system, *Urban Studies*, Vol. 4:61-79.
Gibb, Arthur, 1984, Tertiary urbanization: The agricultural market centre as a consumption-related phenomenon, *Regional Development Dialogue*, Vol. 5, No.1:110-41.
Gubser, Peter, 1983, Jordan: Crossroads of middle eastern events, Westview Press, Boulder, CO.
Hacker, Jane M., 1960, "Modern Amman: A Social Study," Research Papers Series No. 3, Durham: Department of Geography, Durham Colleges in the University of Durham.
Hashemite Kingdom of Jordan, Department of Statistics, 1964, "First Census of Population and Housing," Vols. 1 and 2, Department of Statistics, Amman.
Hashemite Kingdom of Jordan, Department of Statistics, 1953, "1953 Census of Agriculture," Department of Statistics, Amman.
Hashemite Kingdom of Jordan, Department of Statistics 1983, "Agriculture Statistical Yearbook and Agricultural Sample Survey, 1982," Department of Statistics, Amman.
Honey, Rex, Nichols, Stephen, abu Kharma, Suleiman, Khamis, Musa, and Doan, Peter, 1986, Planning regions for regional planning in Jordan's 1986-1990 plan, *in*: J.W. Frazier, B.J. Epstein, and J. F. Langowski, eds., "Papers and Proceedings of Applied Geography Conferences," Vol. 9.
International Bank for Reconstruction and Development, 1957, "The Economic Development of Jordan," Johns Hopkins Press, Baltimore, MD.

International Bank for Reconstruction and Development, 1983, "Jordan, Regional Development," IBRD Mission Report No. 4767-JO, December 7, 1983.

Jones, Barclay, 1985, Interregional relationships in Jordan: Persistence and Change, in: Adnan Hadidi, ed., "Studies in the History and Archaeology of Jordan, II," Department of Antiquities, Amman.

Jones, Barclay, 1986, Urban support for rural development in Kenya, Economic Geography, Vol. 62, No.3:201-14.

Konikoff, Adolph, 1946, "Transjordan: An Economic Survey," Economic Research Institute of the Jewish Agency for Palestine, Jerusalem.

Marshall, John U., 1949, "The Location of Service Towns: An Approach to the Analysis of Central Place Systems," University of Toronto Press, Toronto.

Mathur, Om. P., 1982, "Small Cities and National Development," United Nations Centre for Regional Development, Nagoya, Japan.

Mazur, Michael, 1979, "Economic Growth and Development in Jordan," Croom Helm, London.

Rondinelli, Dennis, 1983, "Secondary Cities in Developing Countries: Policies for Diffusing Urbanization," Sage Publications, Beverly Hills, CA.

Rondinelli, Dennis, 1985, Equity, growth and development: Regional analysis in developing countries, Journal of the American Planning Association, Vol. 51, No.4:434-48.

Stafford, Howard, A., 1963, The functional bases of small towns, Economic Geography, Vol. 39:165-75.

Thomas, Edwin N., 1960, Some comments on the functional base for small Iowa towns, Iowa Business Digest, Vol. 31, No. 2:10-16.

Zahlan, A. B., ed., 1985, "The Agricultural Sector of Jordan: Policy and Systems Studies," Ithaca Press for the Abdul Hameed Shoman Foundation, Amman.

11

The Role of Employment in the Growth of Small and Intermediate-Sized Cities in Egypt

Salah El-Shakhs
Professor of Urban Planning and Policy Development
Rutgers University

INTRODUCTION

The contemporary development of the urban settlement system in Egypt took place under an extraordinary combination of rapid population growth and minimal expansion in its inhabited area. Population density in the Nile Delta (Figure 1) is 1230 persons per square mile which is one of the highest in the world. As a result of extreme crowding in the rural areas, major rural to urban migration flows and rapid urbanization occurred (Table 1). The growth in both rural and urban population, however, continued to be primarily confined to the Nile Valley and its Delta, which constitute a fraction (perhaps 4 or 5 per cent) of the country's total area.

The urban population in Egypt has grown more than 11-fold since the turn of this century. This major shift towards urbanization has taken place with no major additions to the system of settlements. Urban growth has thus been absorbed by existing cities and by rural settlements which became urbanized. Despite the reclassification of a large number of such settlements into towns, however, the bulk of the urban population (over 60 per cent) continued to be concentrated in the two large primate cities of Cairo and Alexandria (Table 2).

This trend towards urban concentration and primacy is nothing new in Egypt. For a brief period between 1960 and 1966, intermediate cities appear to have grown at a faster rate than either Cairo or Alexandria. That period was characterized with an aggressive

Table. 1.   Total Urban and Primate Cities' Population, Egypt: 1897-2000

Year	Total Population	Urban Population	Percentage of Urban Population	Population of Greater Cairo	Population of Greater Alexandria
1897	9715	—	—	985	—
1907	11,287	1930	17.1	1133	—
1917	12,751	2640	20.7	1337	445
1927	14,281	3825	26.9	1676	573
1937	15,933	4493	28.2	2017	686
1947	19,022	6391	33.9	2963	919
1960	26,085	9912	38.0	4820	1516
1966	30,083	12,184	40.5	6113	1803
1976	36,526	16,036	43.9	7300	2383
1986 (est)	48,000	22,500	46.8	12,000	3200
2000 (est)	65,200	35,700	54.8	16,500	5500

Sources: El-Shakhs, 1971; PADCO, 1982; United Nations, 1982; Ibrahim, 1982.

Table. 2.   Distribution of Urban Population and Urban Settlements by size, 1960-76

Settlement size (thousands)	1960		1966		1976	
	Number of Cities	Percentage of Urban Population	Number of Cities	Percentage of Urban Population	Number of Cities	Percentage of Urban Population
20-50	44	15.2	53	18.3	85	20.8
50-100	9	6.7	10	5.7		
100-250	10	18.5	10	13.3	16	17.7
250-500	0	0.0	2	4.5		
500-1000	0	0.0	0	0.0	0	0.0
> 1000	2	56.6	2	57.9	2[1]	61.5
Total	65	100.0	77	100.0	103	100.0

1. Greater Cairo and Alexandria. Giza and Shubra El-Keema are included with Greater Cairo, even though they are separate cities.
Source: El-Shakhs, 1979.

national industrialization program and the construction of the Aswan High Dam, both of which had a favorable impact on secondary cities particularly in Upper Egypt and the Suez Canal regions (El-Shakhs, 1979). This trend was short-lived, and the distribution of urban growth in the 1970s and early 1980s reinforced the trend towards primacy by being skewed in favor of the two largest urban areas. The development of large regional centers to fill the gap in the urban hierarchy has thus far lagged behind national urbanization and industrialization processes.

This trend cannot continue indefinitely. An increase of 15-25 million in the urban population is expected by the end of this century. The capacity of Cairo and Alexandria to absorb such significant population growth is very limited. Even if they grow to their projected sizes of 16 and 5 million respectively by the year 2000, there would remain "some five to 15 million additional urbanites who would have to be absorbed elsewhere within the urban system" (El-Shakhs, 1979:123). This presents a major problem, however, since most of the existing towns and cities originated as marketing and administrative centers in the midst of valuable agricultural land. Their uncontrolled expansion would consume fertile land, a vital resource in short supply, at a much faster rate than any inferior reclaimed land can be added.

To indicate its dissatisfaction with the present trends in population distribution, the government adopted a national urbanization strategy aimed at controlled decentralization of urban growth. Its aim is to divert urban growth away from Cairo but, at the same time, protect arable land from urbanization (PADCO, 1982). The strategy called for decentralization of national government functions, economic and industrial activities, and new housing. This included new efforts to strengthen local governments and to expand the settlement system outside the confines of the Nile Valley and the Delta. President Anwar El Sadat called for the drawing of "a new map for Egypt" which encompasses new desert regions and cities (El Sadat, 1974). The new development regions include: Sinai, the Suez Canal Area, the Red Sea Coast, Lake Nasser, the New Valley, the Qattara Depression, and the North West Coast (Figure 1). The new desert cities are Tenth of Ramadan, Sadat City, and New Amiriyah. The combined total absorption capacity of these new developments is estimated at 5.3 million (El-Shakhs, 1979).

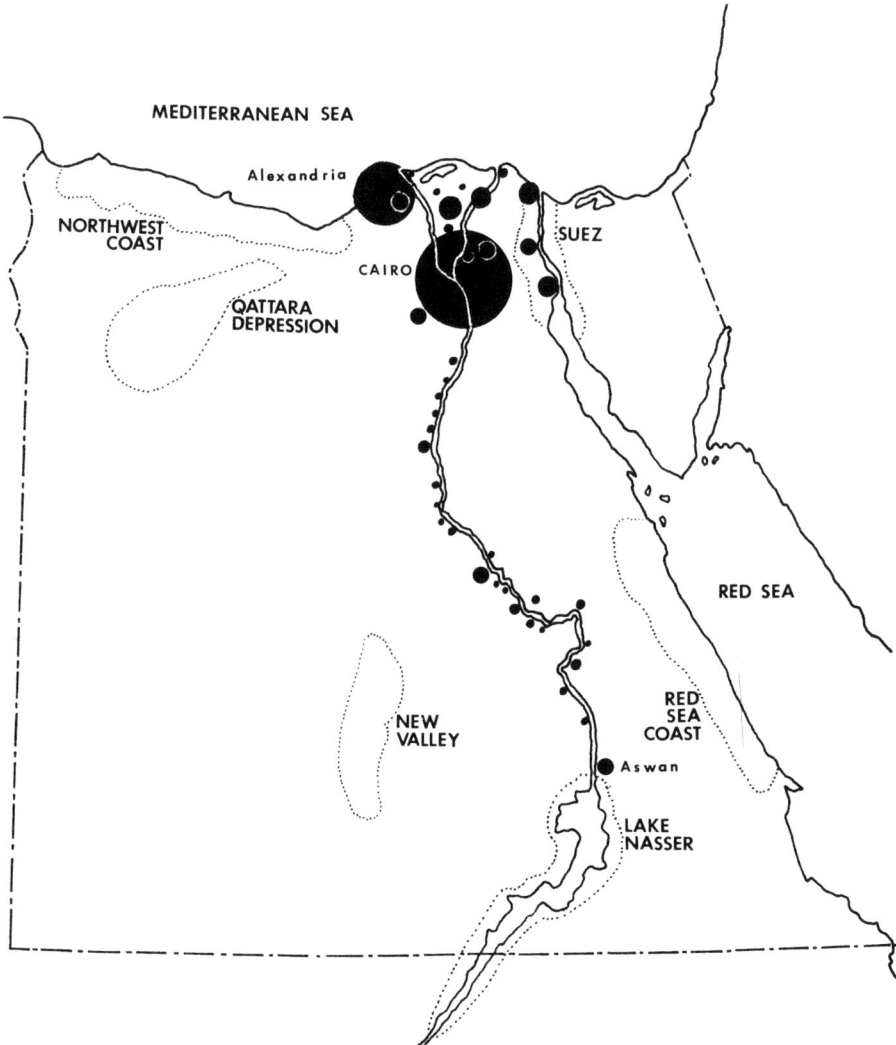

Figure 1.   Major areas of population density in the Nile Delta and
             selected development regions.   (Figure reproduced from
             "Development of Urban Systems in Africa" by R. Obudho
             and S. El-Shakhs (Eds) with the kind permission of
             Praeger.)

THE ROLE OF SECONDARY CITIES

Thus much of the projected urban growth is to be absorbed by small or medium-sized cities, including those in the Suez Canal Region and the new cities under construction in the desert fringes of the Delta. The development of secondary urban settlements is seen as a strategy to concentrate and control urban growth (a type of "concentrated decentralization" strategy), thereby reducing the effects of dispersed urban development on agricultural resources. It is anticipated that such a strategy would create a more balanced urban hierarchy, and thus reduce the pressure on the primate cities, decrease spatial and regional inequalities, enhance rural development and expansion of cultivated areas, and foster embryonic or latent forces of polarization reversal (El-Shakhs, 1983; IURP, 1984).

Defining secondary cities and the factors which contribute to their growth thus became an important issue for research by Egyptian planners. However, defining secondary cities in a functional interaction sense is not an easy task. It would require a significant amount of more detailed information about each city in the system than is readily available. Arbitrary population size distinctions are usually highly inadequate. It was thus decided that we would investigate small- as well as intermediate-sized cities, that is, all secondary cities other than the two primate cities.

THE IMPACT OF EMPLOYMENT

The task of our research was to investigate the relationship between population growth in certain cities, through immigration, and the expansion of formal employment, by sector, as well as improvements in infrastructure. The project has thus far looked only at employment, with the relationship to infrastructure improvement left to a later stage. The aim of the first stage of the research, which is about to be concluded and is being reported here, was to test the prevailing public policy assumption that providing employment opportunities, particularly industrial employment, would attract migrants to small cities. This is important since the location and type of both new employment and infrastructure are determined by the public sector, either through direct investment or manipulation of private capital. An understanding of such relationships would, therefore, lead to a better informed development strategy.

The rates of population growth through migration were computed for the selected cities over the period 1960-76. The growth rates for total employment and for employment by sector in agriculture, industry, construction, and services were also computed. The statistical correlations between rates were then estimated in order to determine the relationship, if any, between types of employment and migration (IURP, 1984)

For the purpose of our analysis, cities were initially classified by population size into two groups: (1) those with 20-100,000; (2) those with 100-500,000. Of the 85 cities in the first group, only those which had positive rates of migration (29 cities) were included, since our primary purpose was to discover factors associated with intermediate city attraction. All 11 cities in the second group, which were identified as regional centers, showed positive migration and were, therefore, included in the analysis. The results of the analysis are shown in Table 3.

As might be expected, a strong correlation was found between the rate of migration and the rates of increase in the active labor force in both city categories. It should be noted, however, that this correlation was much stronger for cities under 100,000 in population. This suggests that the growth of smaller cities is more sensitive to employment opportunities. Both groups exhibited a strong correlation between migration and employment in agriculture, again with a stronger relationship in smaller cities. Small cities in Egypt continue to have a fairly high proportion (over 30 per cent on the average) of their labor force engaged in primary activities particularly agriculture. While the percentage decreases significantly with increasing city size (to a low of 4-5 per cent in cities of over 250,000), the role of agriculture in providing employment for new immigrants to intermediate regional centers seems to be substantial (IURP, 1984).

The correlations with employment in industry were rather weak in both groups, more so in the case of the group of larger cities. Perhaps the most surprising result was the negative relationship between population growth through immigration and increase in employment in the construction industry in either group. This may, at least in part, be explained by the fact that large scale construction relies heavily on non-resident migratory labor, and that residential construction is more a function of savings (particularly from remittances of workers in Arab states) than of population growth. In fact

Table 3.   Correlation Between Rates of Immigration and Rates of Increase in Employment by Sector

Employment Sector	Cities Over 100,000			Cities With 20-100,000		
	r	F		r	F	
Active labor force	0.841	21.7	VH	0.951	276.0	EH
Agriculture	0.708	9.1	H	0.796	50.2	VH
Industry	0.145	0.2	VL	0.248	1.9	L
Construction	-0.724	9.9	-H	-0.316	3.2	-L
Services	0.394	1.7	L	0.683	25.2	H

r = Correlation coefficient; F = Fisher value;
H = high, V = very, E = extremely, L = low.
Source: IURP (1984).

it is frequently observed that relatively high levels of construction activity exist in secondary cities even if their populations are growing at lower than average rates.

It thus appears that one can tentatively conclude on the basis of this research so far that increases in employment or in construction activity are not strongly related to attracting migrants to specific cities. Other factors such as location, function, amenities, level of services, and quality of life whether real or perceived, may be more important as attraction forces than either employment or housing. The second stage of this research aims at classifying Egyptian cities on the basis of such variables, and looking at their relationship to immigration.

Acknowledgement

This paper is based primarily on a research project on Employment and Infrastructure Planning for Secondary Urban Settlements in which the author was a co-investigator with Drs M. Yousry, T. El-Sadek, and A. Barrada of the Institute of Urban and Regional Planning, Cairo University. It was funded by grant No. 830702 from the

Foreign Relations Co-ordination Unit of the Supreme Council of Universities in Egypt as part of the University Linkages Project.

REFERENCES

El Sadat, Anwar, 1974, "October Working Paper," Government Press, Cairo.

El-Shakhs, Salah, 1979, Urbanization in Egypt: National imperatives and new directions, in: R. Obudho and S. El-Shakhs, eds., "Development of Urban Systems in Africa," Praeger:116-31, New York, NY.

El-Shakhs, Salah, 1983, The role of intermediate cities in national development: Research Issues, in: M. Chatterji et al, "Spatial Environmental and Resource Policy in the Developing Countries," Gower, Hampshire.

Institute of Urban and Regional Planning (IURP), 1983, "Employment and Infrastructure Planning for Secondary Urban Settlements: Report #1, Detailed Plan of Action," IURP, Cairo University, Cairo.

Institute of Urban and Regional Planning (IURP), 1984, "Employment and Infrastructure Planning for Secondary Urban Settlements: Report #2, Employment and Population Growth," IURP, Cairo University, Cairo.

Institute of Urban and Regional Planning (IURP) 1984, "Employment and Infrastructure Planning for Secondary Urban Settlements: Report #3, Employment, Size and Growth," IURP, Cairo University, Cairo.

Institute of Urban and Regional Planning (IURP), 1985, "Employment and Infrastructure Planning for Secondary Urban Settlements: Report #6, Functional Classification of Egyptian Towns," IURP, Cairo University, Cairo.

Mathur, Om. P., 1982, The role of small cities in national development re-examined, in: O. P. Mathur, ed., "Small Cities and National Development," United Nations Centre for Regional Development, Nagoya, Japan.

PADCO Inc., 1982, "National Urban Policy Study," Ministry of Development, new Communities and land Reclamation, February, 1982, Cairo.

Rondinelli, Dennis A., 1982, Intermediate cities in developing countries: A Comparative analysis of their demographic, social and economic characteristics, Third World Planning Review, Vol. 4:357-86.

12

Secondary Cities in Kenya: Problems of Decentralization, Municipal Finance and Urban Infrastructure

David B. Lewis
Associate Professor, Department of City and Regional Planning
Cornell University

CONTEXT

Located at the equator on the coast of the Indian ocean in East Africa, Kenya has a population of 20 million people. It has a very high growth rate at approximately 4 per cent per year. Most of the population, 85 per cent, lives in rural areas and is engaged in agriculture. Only 20 per cent of the land area of the country is suitable for agriculture, and that is already saturated at the maximum density that can be supported. There is very limited opportunity for horizontal expansion of the rural population into new areas. The labor force of the country will double by the end of the century. This is not a statistical projection but a demographic fact: the new entrants to the labor force at that date have already been born and are children with names. The urban population will increase from three million in 1984 to nine to ten million by the year 2000. This will mean that the urban population will increase as a percentage of total population from 15 per cent to 25 or 30 per cent of the total. Even then, the increasing pressure of rural density will present serious problems.

PROBLEM

The government faces a two-fold problem. First, there is an urgent need to stimulate the growth of economically productive employment opportunities to absorb the expanding labor force. Second, there is a need to facilitate the development of productive urban structure to meet the requirements of the growing urban population and also to help to relieve the pressure of excessive rural densities.

Employment

It will be necessary to at least double the number of job opportunities in the next dozen years. This will take care of only the increment in the labor force so present problems will not worsen. To deal with unemployment or underemployment will require even more jobs. The agricultural sector does not provide a solution. It cannot productively absorb the projected increase in the labor force. The available agricultural land is already being subdivided into units too small to support a family adequately. Furthermore, improving agricultural productivity to provide adequate incomes for agriculturalists requires increasing the capital intensity of agricultural production technologies. The ability of agriculture to absorb labor will be even further reduced.

It is prohibitively expensive to create a significant number of new job opportunities in the capital intensive industrial sector. High productivity manufacturing must expand for the economic growth and well-being of the national economy. However, the new jobs created will not begin to absorb the increase in the labor force.

Urban Structure

The basic need is to encourage the growth and development of more small and medium sized urban places. Urban services are difficult to obtain at any distance from the larger urban centers because of the absence of this type of urban place in extensive agricultural regions. Economies of urbanization or agglomeration need to be exploited. The costs of facilities and infrastructure could then be shared by economic and administrative activities. Economies of scale need to be exploited by creating larger markets for enterprises through the expansion of modest-sized urban places.

The increase of small and medium sized urban places would help to avoid the problems of excessive concentration. Economic activity and the availability of goods and services is already somewhat concentrated in a few major centers. If the increment in urban population is primarily absorbed in these centers, real problems of concentration would result.

More small and medium sized places would address the need to foster backwards and forwards linkages with agriculture. For agriculture to become more productive, it needs increasing intermediate goods inputs from the urban sector. Such places would provide readily accessible market points and locations for primary processing activities of agricultural products. They would provide in that way additional incentive to increase agricultural productivity.

There is a problem of resource allocation between the most productive areas and arid and semi-arid regions. Small and medium sized urban places could help to meet the need to address this problem. The less productive regions are geographically very extensive. Populations living there are often extremely remote from governmental services and urban goods and services of all kinds.

SOLUTION

Rural Trade and Production Centers

The Government of Kenya has adopted a new policy to emphasize the growth and development of rural trade and production centers (RTPCs). Sessional Paper No. 1 of 1986 elaborates this strategy enunciated in the 1983-88 National Plan. It is proposed to assist approximately 200 RTPCs. By the year 2000, seventy will be aided at the extent of $1,000,000. The government assistance is to be used primarily to finance critically needed infrastructure. To accomplish this plan, the administrative capacity of local authorities will be strengthened.

Non-farm employment opportunities in these centers will be promoted. The emphasis will be on expanding the role of the informal sector. Infrastructure will be built in support of productive activities. Linkages between secondary towns are to be strengthened so that the urban system operates more as a hierarchical network than a monocentric system. Scarce government resources are to be allocated to small urban centers with highest potential for growth.

To carry out this plan a number of issues must be addressed. Decision-making must be decentralized in the first place. There is at present a strong tendency for authority over even very local issues to be retained at the ministerial level in the central government. As a consequence, experience in administering and making decisions is lacking at the local level, and the resulting lack of capability further reinforces centralization. A second issue that must be confronted is that of municipal finance. Local governments have extremely limited powers and mechanisms for generating revenues. Consequently they are heavily dependent on the limited budgets of central ministries. A third issue involves the lack of urban infrastructure. With limited financial resources at the municipal level and limited authority, very little has been developed in terms of the basic urban infrastructure that is necessary to support the existence of productive activities in small centers.

Decentralization

There is a real concern on the part of the central government with relinquishing political power. Control of governmental operations often seems tenuous enough without releasing authority. Administrative capacity, trained and experienced personnel, ease of supervision of performance and execution seem more efficiently manageable when centralized than dispersed.

Local authorities are currently administered by the Ministry of Local Government. They are dependent on the Ministry for annual allocations from the central government that provide the bulk of their revenues. Local authorities have almost no autonomy and can take few actions to meet their own problems except to request assistance from the central government.

To provide greater autonomy at the local level, the government adopted a policy for delegating local development planning and management responsibilities to districts. This was implemented through the District Focus for Rural Development program. While this was a major step toward decentralization, it was not intended to address the development administration needs of urban areas. The composition of the District Development Councils was not designed to deal with urban problems. The program was oriented more toward problem-solving rather than a strategic perspective. The Councils were better designed to cope with current and immediate issues than to consider promoting development in a longer range sense.

Municipal Finance

Local governments in Kenya have severely limited authority to raise revenues. Fees for services, licenses, and similar instruments are the chief sources. Powers of taxation are extremely limited. These sources of revenue yield minimal resources which can be used for discretionary expenditures. Not only are local governments dependent on annual allocations from the central government to meet their financial needs, even their ability to borrow funds against anticipated income flows is constrained, and they must seek loans by application to the central government. There are also limitations on their obtaining resources to fund critical recurrent expenditures or to undertake capital expenditure projects.

The central government is highly dependent for its revenues on indirect taxes, primarily import duties. Such taxes can yield only quite limited revenues under the best of circumstances. In the case of Kenya this source of revenue is particularly unsatisfactory. Imports are limited by shortages of foreign exchange. The government is also trying to reduce the dependency of the economy on imports. As a consequence, the central government simply does not have the resources to make adequate allocations to local governments and meet the needs of its own programs as well.

The informal sector is the most important economic activity in small and medium sized urban centers. However, it is less easily taxable than large scale agriculture and formal sector activities. It is not an extremely promising source of local revenues, but it is critically important to the economic function of these centers. There is a need to find ways to stimulate the development of the informal sector. Actions which increase agricultural productivity, and consequently farm incomes, will create increasing demand for the goods and services produced by the urban informal sector. A revision of the tariff structure could reduce the cost of informal sector inputs, either raw materials, equipment, intermediate goods, or other articles. The government could take actions which would improve the investment climate in the informal sector and create incentives which would mobilize private capital.

In sum, the problems of municipal finance are two-fold. There is a need to devise new sources of local revenues, and to reduce impediments to the growth of the informal sector which is the major economic activity in smaller centers.

Urban Infrastructure

Most of the small urban places have recently grown from traditional
hamlets or villages. They have virtually no urban infrastructure and
no accumulation from past periods of investment and construction.
The rapidly growing populations of these places is generating a
greater demand for infrastructure. As population increases, the
needs for sanitary water supplies, sanitation systems, control of rain
water run-off, streets and walkways passable in all weather, not to
mention a host of other items, become critical.

Interurban linkages must be improved also. As places become
larger there is increasing need for interaction between them. Inter-
dependencies grow and require better transportation facilities and
means of communication. Urban infrastructure needs extend beyond
urban boundaries to include the kinds of facilities that permit urban
places to function.

The urban infrastructure needs include human resources also.
Trained and skilled personnel who can plan, construct, manage, and
operate infrastructure systems musᵗ be available for the systems to
function. Management practices and organizational efficiency are
especially vital where resources are scarce.

A number of policy emphases are suggested by these considerations for urban infrastructure development. (1) Engineering standards must be adjusted to make them more appropriate to current
realities. (2) By concentrating on a single project, or small number of
them, maximum economic utility can be realized. (3) It is important
to charge fair market prices for urban services, such as collecting
user fees, to create flows of funds to meet recurring expenses. (4)
Since general revenue sources are clearly inadequate to provide the
types and levels of urban services that are demanded, it is vital to
select investments that will yield high rates of financial returns. This
will generate revenues that can provide subventions for other
services that are not self-financing and will also create funds that can
be used for other investments. (5) Privatization of services should be
explored to the extent possible. Many urban services can be provided by the private sector through franchises, contracts, and similar
instruments. This can release government resources to provide other
service needs. (6) There is substantial potential for mobilizing nongovernmental resources to develop needed infrastructure.
Incentives and regulations can induce private response to needs for

physical facilities and provision of services. Numerous examples can be cited in both the past and the present in which this approach was extremely successful. After all, a great deal of what exists in the way of infrastructure antedates the central government.

CONCLUSIONS

The development of secondary cities is a critical issue in Kenya. They are necessary to absorb much of the increase in urban population that is imminent. They provide a means of reducing the pressures of increasing rural densities. The enormous increase in the labor force that will occur in the next decade and a half must be largely employed in the informal sector. Small and medium sized urban centers provide locations and functions for these activities. Such places can make urban goods and services more accessible to rural populations increasing their productivity and well-being.

Impediments to the development of secondary cities must be reduced. The central government must be willing to decentralize to a larger extent, even though this means relinquishing some degree of political power and authority. A major impediment to the growth of urban places is the problem of municipal finance. Local governments must find expanded means for raising funds to meet the needs of increased services. A further impediment is the lack of urban infra-structure and the need to develop human resources, physical facilities, and service providing organizations.

All of these issues are interrelated. The growth of the popula-tion and labor force require greater urbanization and urban employ-ment opportunities. Further concentration would be counterproduc-tive, and dispersion of urban places would serve agricultural populations and remoter areas well. However, secondary cities cannot develop to meet these needs unless solutions are found to the problems of decentralization, municipal finance, and the provision of urban infrastructure.

Changing Priorities for Consultants, Donors, Clients and Counterparts: Views and Experiences

13

Egypt: Profusion of Plans, Poverty of Programs — A Consultant's Experience

Malcolm D. MacNair
Development Consultant, Pt Hasfarm Dian Konsultan, Jakarta, Indonesia

This paper provides a brief history and appraisal of development planning in Egypt with emphasis on activities since the War of 1973 and the infusion of foreign, largely Arab and Western supported, technical assistance. More than 100 major plans and studies prepared in the period 1973-85 covered the entire country. These studies varied from very inclusive regional plans to designs in construction detail for new communities scaled at a half million persons each. Drawing upon the best of British, Dutch, French, Canadian, American, Japanese, German, Yugoslav, Italian, Swedish and Norwegian consultants, along with their Egyptian counterparts, many of these plans were on the technical frontier of planning analysis and method. Yet, few elements of these plans were either carried out or welded into a national program for action. This state of affairs in part occurred due to: problems similar to those in other developing countries including the misuse of human and other resources; the perpetuation of early models of administration which are insufficient to later tasks; and the emphasis on technical assistance projects instead of an emphasis on technical assistance processes. This paper proposes a corrective direction in the development and foreign aid process and the role of expatriate professionals.

BACKGROUND AND MOTIVES GUIDING PLANNING

Hydroelectricity, Heavy Industry, And Reclamation Agriculture:
The First Planning Phase

With the revolution of 1952, Egypt, as with other then emerging nations, sought rapid development following the industrial growth patterns of the Soviet Union and the United States through the exploitation of natural resources, especially irrigation and energy, and the development of steel and other heavy industries. This direction, involving projects beyond the financial abilities of Egypt alone, focussed on construction of the High Dam near Aswan and related projects to extend agriculture by land reclamation, to increase the intensity of agriculture through flood control and triple cropping permitted by those controls and perennial irrigation, and in a later rationalization, to produce aluminium and fertilizer tied to the production of electricity.

The idea of a new dam at Aswan and the creation of the world's largest man-made lake was neither revolutionary affectation nor pharaonic ostentation. Egyptian population, as with other developing nations, was increasing from declining death rates and constant, sometimes increasing, birth rates associated with improvements in the general standard of living (Note 1). An earlier dam at Aswan, built by the British and increased in height in 1912 and again in 1933, brought a million feddans of new agricultural land into production and increased the amount of arable land to about three per cent of the total land area (Note 2). When the old Aswan dam was first built, the total population was about 12 million persons. In 1952-55 when West German experts were studying the possibility of a new dam, the total population was about 25 million persons, or only twice that of Cairo today (Note 3). The High Dam would add, it was projected, another 1.25 million feddans of agricultural land, increase arable land to about four per cent of the national area, and provide the nation's needs in agriculture and electricity for years to come (Europa Publications, 1984).

With a change of commitment and the refusal by the United States to help build the High Dam, and events following the nationalization of the Suez Canal in 1956, construction of the dam and national development took place through the 1960s under the influence of Russian assistance and administrative practice. The resulting administrative system, which borrowed some practices and influenced subsequent administrations, was less socialist than it was uniquely Egyptian, and especially, Nasserist (Nutting, 1972).

The Economic Agency, the central planning authority created in 1957, produced the first Five Year Industrial Plan in 1958. Concerned with industrial investment and economic recovery from the Suez Canal, this plan did not attempt to integrate plans for infrastructure supporting development or physical planning related to reforms (Note 4).

Regional Planning In the First Planning Phase

Regional planning began during the 1960s with Russian advisors and the belief that natural resources in the Western Desert (as elsewhere) were exploitable without cost except for cheap Egyptian labor. Assessments of iron ore and phosphate deposits in the Western Desert, and the results of groundwater studies of the Nubian Aquifer west of the Nile showing an underground sea of water, provided the idea for a bold thrust in the development of the New Valley. Besides the old valley of the Nile, then restricting national development ambitions to a 1200 kilometer long, green strip through the desert, 15 kilometers across at its widest, the government would develop a chain of oases in the Western Desert as a region to absorb the still growing population. This new region would engage in agriculture to reduce the growing threat of a need to import food and in mineral extraction to raise the wealth of the nation and provide an expanded industrial base.

Like the High Dam, New Valley was a stroke that captured the popular mind. Although there was little data to support the claim quickly assembled, or modified to political hope, the government would develop 500,000 feddans for agriculture and provide settlement for about as many people (Note 5). Studied and constructed in haste, soils proved, contrary to plan, inadequate. Water conduits and other infrastructure decayed through poor design, construction, repair, maintenance, or all these failings together. New settlements were abandoned. In some oases, whose populations had been in ecological balance for 200 years, farmland was lost almost immediately as deep well drilling dropped the aquifer to depths beyond traditional wells. In others, existing lands were lost slowly as increased irrigation made heavy, poorly drained soils too salty, or soggy with elevated water tables, for plant growth.

The Plan for Greater Cairo, 1970: The Second Planning Phase

Begun in 1960 and completed in 1970, the High Dam proved successful in many of its planned objectives, especially in flood control and
in the production of electricity.   By 1974, revenues attributed to the
dam paid for its construction.   Urbanization, however, was rapid,
increasing from about 20 per cent in 1917 to about 50 per cent by
1977.   With a consequent loss of arable land, especially in the 1950s
and 1960s, there was an increasing recognition by 1970 that
continued loss through urbanization could cancel any gains in arable
land made by construction of the High Dam and land reclamation
(Note 6).

In the late 1960s, those concerned with both national and urban
development could envision five ways to address the land-population
problems Egypt faced:

1.   Extend the land under cultivation;

2.   Intensify production of existing arable land and preserve arable
land for human crop production;

3.   Absorb   excess   population   in   non-rural   pursuits   (mining,
industry, tourism) on non-arable land — in new towns, urban in-fill,
and so on;

4.   Gain population control; and

5.   Promote emigration (or a substitute, such as the export of
workers for salary remittances).

The Metropolitan Planning Commission for Greater Cairo (Note 7)
aimed at the third of these ways and developed a plan to direct rapid
Cairo growth out of the valley to desert land.   There would be a
system of satellite cities (Note 8), divestment of new employment
and other attractors of population to these regional subcenters, and —
much like Paris and Moscow — an outer ring road to contain, in
theory, central city growth.   This plan seriously discussed the
promotion of emigration and emphasized the need to preserve
agricultural land.

Modern Cairo, originally constructed with a much smaller population and more gradual growth in mind, was built to accommodate

two million persons.   Cairo in 1970 had four million or more persons.
With various overloaded and seldom maintained channels, the city
slowed in its ability to function.   Sewers broke and sewage backed
into the streets.   Water pressure in the mains dropped and groundwater infiltrated the residential piped system.   Traffic was congested,
if not grid-locked, during working hours.   There was, in addition,
unnecessary trip generation and cross-flows of workers living
opposite one another's employment (Note 9), solid waste accumulation, and air pollution.   Modern ills and the press of too many people
too soon threatened to overcome an old city.   Along with correcting
these difficulties, the plan called for reducing population densities to
gain relief not only from the overcrowding caused by rapid urbanization, but also by squatters displaced from the Canal Zone cities by
war.   The plan renewed the mission to preserve agricultural land by
moving into the desert and completed the ideological template for
Egyptian planning.

The 1973 October War, The Suez Canal Zone Reconstruction,
And After: The Third Planning Phase

The war of 1967 closed the Suez Canal and caused the evacuation of
most inhabitants of the region to other urban centers, principally
Cairo.   Armistice negotiations, following the war of 1973 and the first
oil crisis, anticipated the clearance and re-opening of the canal by
mid-1975.   In 1974, the Ministry of Development (then Housing and
Reconstruction) obtained authority to restore normality to the region
and to plan its development.   The Minister, a confidant of President
Sadat and head of the construction company that had built the High
Dam, created the Advisory Committee for Reconstruction (ACR) to
oversee the planning for the region.   He created a Central
Organization for Reconstruction to carry out rebuilding and the
development recommended by the ACR.   The Minister provided the
services of the General Organization for Physical Planning (GOPP) to
the ACR and sought United Nations Development Program (UNDP) aid
for the foreign exchange costs of additional technical assistance.

FOREIGN ASSISTANCE IN PLANNING

The UNDP at the beginning of 1975 provided the services of a development advisory group (Note 10), which, jointly with the GOPP and
an American consulting firm (Note 11) serving as in-house advisors
to the ACR, prepared the Suez Canal Regional Plan in late 1976.   The

regional plan contained subregional plans, or area masterplans, for the three canal cities, Port Said, Ismailia and Suez, and proposals for a larger region surrounding the Canal Zone. At the request of Egypt in mid-1975, the UNDP approved the establishment of a cost-sharing arrangement under which member countries contributed to a series of follow-on studies that the regional plan identified.

As a major focus for international donor activity and for the coordination of all these activities, the ACR extended its realm to the development of Egypt beyond the canal region. Until 1985, it was the center for development planning in Egypt. The Appendix lists some of the most important studies carried out in the period 1973-85, either with the policy determination, guidance, or review of the ACR.

With Law 43 of 1974, the government declared an open-door policy allowing and attracting foreign investment. Until 1978 this was a period of domestic emphasis in which there was a coordination of economic policy and development planning. The Government greatly advanced the reconstruction and development of the Suez Canal region.

The canal cities were rebuilt. Improved roads linked them and Cairo. New electric power stations, bridges and housing served them. The canal, cleared, was being widened and a tunnel beneath it was under construction to re-link Africa and Asia. The nation's first four-lane highway connected the Canal Zone and Cairo. The Government developed a free zone in Port Said and industrial and oil-harboring facilities in Suez. Agricultural lands were re-settled and extended. The railroad between Port Said and Ismailia re-opened. Studies called for under the 1976 regional plan were underway.

Elsewhere, new land reclamation started near Alexandria and the Fayum. The iron and steel complex at Helwan was expanded and, among other materials and industrial projects, the Nag Hammaddi aluminium plant inaugurated. Hotel construction began in Cairo and Luxor as part of the revival of the tourism industry. Yet despite this considerable activity, among expatriate experts, aid donor professionals, and Egyptian technicians themselves, especially after 1980, there was frustration in the belief that in Egypt nothing ever goes or gets done as planned. Such views without statistical account can be attributed to the high hopes of those who believe in the policies they recommend, in the potentials of their plans, in the quality of their technical or professional analyses, or perhaps, in the amount of foreign aid granted.

APPRAISAL OF PLANNING

Often normal sequences of procedure cannot be accelerated and what appears as delay is an underestimation of the time actually required to do the tasks; for example, the project identification study, pre-feasibility study, and feasibility study sequence followed by most donor agencies. In Egypt, with so many plans and with so few resources, a timely execution of plans was, perhaps, impossible. Data, too, are often of questionable quality and delay or added study to resolve an empirical question is valid when involving large sums of money or the possibility of irreversible, adverse results.

Whatever of these or other considerations, the consensus among expatriate development professionals remained: the prospects for administrative improvement in the near term were discouraging. The Government implemented very few of the projects identified. For example, in the detailed, follow-on plans for industry, agriculture, tourism, solid waste, and the development of Suez Governorate in the canal region, no program existed to order the implementation of projects among plans and no authority or deliberately designed process for integrated plans (Note 12).

Organization

Physical development planning and national economic planning are separate activities in different ministries without any device to coor-dinate the two activities. There is a national budget and budgeting process in the Five Year Plan, assembled by the Ministry of Planning and the Ministry of Finance. However, the process only treats budget amounts and does not combine related projects originating from different units in the sense of a plan, or a sense of relationship to a greater purpose than those of the separate projects alone. For exam-ple, the road construction projects of the Ministry of Transport, those of the Ministry of Development and those of the various governorates were treated as amounts in separate unit budgets, not in an antici-patory way and as possibly related — possibly duplicated — road con-struction projects (Note 13). Economic planning consists of selecting investments to achieve a set rate of growth — about five per cent in the early 1970s, about ten per cent in the early 1980s — called for in the national Five Year Plan.

By a 1979 law, the Ministry of Planning (MOP) was to extend this process to a set of regions created by the law and to the governorates comprising them. The centrally developed plan for each governorate (Note 14) and region provided general budget ceilings, or targets, based upon past spending, population ratios, or other rules for the distribution of expected funds for local government investment. The MOP regional staff adjusted and combined the investments scheduled at the governorate level into a regional budget. The national investment budget followed, supposedly, as the alignment of budgets among regions after adjustments back and forth across ministries and other units of government.

The activities of the MOP in economic planning for regional development were unrelated to the regional planning activities of the Ministry of Development (MOD); a frequent state of affairs as new parliamentary laws and presidential decrees often did not consider or modify, existing law. By the legislation creating it in 1974, the MOD had a regional development mission similar to the one later granted to the MOP, except theoretically it was limited to the territory outside the Nile Valley and to an indirect budgeting role.

Accepting the notion in the abstract that the Egyptian government was too centralized, Sadat, and later President Mubarak, proclaimed decentralization as "central decision for local implementation". In that logic, the responsibility was the governors'. The resources for local implementation, nonetheless, were lacking, particularly in skilled personnel and local powers to raise revenues. Since the governors had not engaged in determining the projects and there were insufficient technical staff to explain them, the import of projects were at best ambiguously understood, and therefore, ambiguously pursued.

Political Heritage

The substitution of the military for middle class government officials not only minimized democratic review of government policies, centralized command, and emphasized a more unquestioning deference by inferiors to superiors in public administration, but also established practices stemming from a lack of expertise in substance and in governance that permeated the civilian government. Often without a theory of practice or aided by independent points of view, decision makers sought the quick solution and the appearance of command. They moved from one solution to another, avoiding the appearance of failure, but relinquishing a reasoned objective.   At times a new crisis could supplant the urgency of the old.   Invariably, if it was a public matter, the public was assured that those at the top were giving their full attention and that those best able were in committee forging a solution.

For example, the government sought to provide every person housing, but set standards too high to be able to afford to do so.   The government, nonetheless, rejected means to obtain housing for all, because those in positions of decision believed that somehow these means were insufficient and, thus, inappropriate.

The Sadat City Development Authority built core housing to an almost completed residence and discovered, comparing the cost with regular government-provided housing, that core housing was too expensive to be a widespread housing solution.   Although the inhabitants of El Hekr Project in Ismailia recognized it as a success, as expatriate consultants did also, Egyptian housing officials viewed El Hekr, and sites and services in general, as a failed solution to the housing problem, because the housing obtained was incomplete. Consequently, the government complained about an informal housing sector and continued to build, at greater cost but in insufficient supply to meet demand, public housing or, in Tenth of Ramada, faulty prefabricated, concrete housing with jointing gaps.

The planning rationale for Fifteenth of May, the satellite town built to reduce the cross-traffic commuting of steel workers from Cairo to Helwan, illustrates the general problem of exaggerated standards in Egyptian planning and plan execution.   Initially to have been financed in part by Saudi Arabia, early British garden city features strongly influenced the design concept.   The designers planned one-third of the town area as open space, which is difficult usually to maintain in an Egyptian context.   Although the land was cheap, the expansive design added to the cost of infrastructure.   Situated on a limestone plateau with little soil cover, the realization of the garden city plan required blasting holes into the stone surface and using agricultural soils in order to plant trees at about US$150 a tree.

More importantly, there were no housing market studies, and consistent with the design, construction standards were set too high for affordable housing on a steel worker's wage. Sales for the first phase of the new town were slow with consequent added financing costs, speculation and eventual purchase by persons working in Cairo. Contrary to the purpose for building this new town, traffic congestion from cross-commuting continued and increased.

REFLECTIONS ON EXPERIENCE

Technical Assistance Processes, Not Projects Alone

If there is a profusion of plans in Egypt and a poverty of result, some contribution to this state of affairs should be laid to the structure of the foreign aid process and the drive for a concrete product directly identified with such aid. Aid agencies are accountable to their governing body and must show results for their efforts.

There has to be a definite product, although planning should be a process to anticipate changing circumstance and opportunity, in a definite period of time. As a result, there are tight schedules for the use of experts and little — chaotically assigned — time for reflection, beyond a specific assignment, and for the training of counterparts, which is usually written into aid agency contracts as a standard, but often ignored, item. This is unfortunate, because reflective and educational processes are essential to the development process.

Equally unfortunate, the recipients of the studies often confused studies of process — plans — with studies purely of projects — feasibility studies. In Egypt, this led to the rejection of two important studies, the second New Valley regional plan and the National Urban Policy Study, because of the drive for a completed study within a set time and a set expectation of a product, or a final solution. Both studies had models with which to explore policy options, the one relating groundwater and agriculture systems and the other relating the national economy and variations in the costs of urbanization. The contractual pressure did not permit the understanding that the models were to be used in an on-going way to test policy variations and changing circumstances.

Contractual time and scheduling disallowed training in the Ministry for the use of the models during the study period. Those who funded the study did not anticipate providing financing for training for model use after the completion of the original study

periods.   The studies were rejected largely because, without policy
explorations, the client believed that there were no other options
than those presented by the consultant.   The National Urban Policy
study was rejected, for instance, because it found the new cities too
expensive in relation to the ability of the national economy to pay for
them and placed the Minister in a position of political and public dis-
comfiture.   Possibly, rejection of this study could have been avoided
if extended exploration of policies by use of the model with the
Minister and his staff had been anticipated in the original study
design.

New Role for Foreign Experts

The many plans of the MOD can be consolidated into a plan and
development strategy as a technical matter.   Incorporated in the Five
Year Plan to provide a developmental direction to a budgeting
process, there could be a gain, not only of a logical culmination to a
great effort by donor agencies, but also of a clearer direction for the
reassignment of excess government employees linked to national
development.

    To achieve this, foreign experts will be needed, but in ways
different from the past under the pressure and drive to assemble a
product.   Rather than a counterpart arrangement, expatriates and
Egyptians would be integrated into a team where all will have
assignments and accountability under a unified management.   This is
a collegial system in which the accomplishment of a task usually
emphasized for the foreign expert to perform will be mutually
shared, with the difference that if training is required, it has to be
provided as part of the joint effort of the work of the team.

    There would be a mutual learning by this process as connections
are made by the team throughout the government to establish
structures for national development planning and budgeting linked to
national economic planning.   Fewer expatriate staff would be
involved in the work, but for a longer term.   Along with the introduction of methods of financial evaluation, routine data collection, text-
based data referral systems of plans and projects, organizational
analyses and the like, this development-as-learning process will aim
at forming programs to isolate, analyze, and restructure government
and its internal cultures.   Falling between ministries and between
governorates, this program to form a national development program
should be located in the Prime Minister's Office.   There, planning

could greatly help local development efforts in a unified drive to program plans and to carry out projects for a national advance. With limited land and water, the remaining and most important, developable resource that Egypt has is its people.

Acknowledgements

This paper reflects the author's experience as an Advisor to the Egyptian Ministry of Development, New Communities and Land Reclamation from 1980 to 1985. Much of the material drawn upon is in sources and files left in Egypt. Other material stems from conversations with colleagues and direct experience, particularly in the author's participation in the second New Valley Regional Development Plan, the Sinai Development Study, among other development studies, and efforts to institute programs of organizational and administrative change in the Ministry, including training.

REFERENCES

Europa Publications, 1984, "The Middle East and Africa 1984-5," Thirty-first edition, Europa Publications, London.
Nutting, Anthony, 1972, "Nasser," EP Dutton:299, New York, NY. (Nutting's book is a particularly good one in its inter-relation and interpretation of Egyptian external and internal political events. Nutting had a relationship with Nasser both as a British diplomat and, after resigning over the Suez invasion of 1956, as a journalist and friend.)

NOTES

1.   Deaths in 1912 were, for example, about 29 to 30 per 1000 persons.   In 1977 (the year of the last reported census by CAPMAS, the Egyptian Central Agency for Population, Mobilization and Statistics) death rates were about 15 to 16 per 1000.   Birth rates in 1977 were about 37 per 1000, a slight decline from the estimated rates for 1912.   While the rate of population increase in 1912 was just over one per cent, it was about 2.5 to 2.6 per cent in 1976.   The present rate of increase is about 2.2 to 2.4 per cent, even with relatively high infant mortality rates (Egypt has an infant mortality rate now of about 100 to 120 infant deaths per 1000 births per years, according to an estimate by USAID in the "Child Survival"

Project Paper, Cairo: USAID, circa 1986.   See also World Bank statistics).

2.   A feddan is slightly more than an acre, or 1.038 acre (4201 square meters).   The usable area of Egypt is about 38,8000 square kilometers (about 15,000 square miles).

3.   The present national population is about 45-50 million persons. Current projections advance this number to 75-80 million by the year 2000.   USAID projects a population of 20-30 million persons for Cairo alone in the year 2025.

4.   Physical planning, contained in the then Ministry of Planning and Local Government, was involved in planning and carrying out housing, urban renewal, and slum clearance projects.   These concerns at the time were also those of the United States.

5.   Average farm size is five feddans.   Average family size is five persons.

6.   These are gross estimates and there are other contributors to the loss of arable land.   Kishk estimates that about eight per cent of the arable land, or its equivalent in lowered production, is lost to desertification along the valley edge.   He estimates that brick making removed another 120 square kilometers.   He states that 80-200 square kilometers per year is being lost by the expansion of rural settlements and urbanization.   Greatly affecting productivity, an estimated one-third to one-half of agricultural land may be salinized and required drainage at an average cost of US$200 per hectare (see Mohamed Atif Kishk, "Land Degradation in the Nile Valley", AMBIO, Vol. XV, No 4, 1986).   The National Urban Policy Study, completed for the Ministry of Development in 1982 and partially financed by USAID, used Landsat data to estimate that 120 square kilometers of arable land per year were being lost by the expansion of settlement, especially recently, in the Nile Delta. A USAID-Cairo study of Egyptian agriculture in 1976, estimated that produce could be increased about 50 per cent in Egypt by better harvesting, packaging and transportation of produce to reduce food wastage.   This finding remains valid. Mechanization, by reducing the amount of land needed for animal feed, could gain 30 per cent more land for human food crops.   Further mechanization, of course, increases urbanization.   The degree of agricultural mechanization, therefore, is a choice for an urban policy, but one complicated in Egypt by small and scattered small holdings.

7. The Commission represents three governorates. This organization later became part of the Ministry of Development and, under the name the General Organization for Physical Planning, is now responsible for the planning of all cities and towns in Egypt.

8. El Obour to the southeast of Cairo, El Amal to the east, Sixth of October to the southwest, and Fifteenth of May to the south just southeast of Helwan.

9. Fifteenth of May illustrates this. The steel mill initiated in 1968 at Helwan drew its workers from Cairo while persons living in Helwan, formerly a small resort and agricultural service town, worked in Cairo. Cross-commuting traffic, especially with a mix of truck, bicycles, animal cart, donkey and camel (outlawed from entering Cairo in 1970) strained land capacities in both directions on the single riverine road connecting the two cities.

10. Comprised mostly of faculty from the University of Glasgow.

11. Tippetts, Abbett, McCarthy and Stratton (TAMS).

12. This is true for most of the regional plans listed in the Appendix. The three free standing, new cities are under construction with Tenth of Ramadan, the most advanced in construction, dependent on workers commuting from Cairo. Much of the canal cities projects are built and planning units are established in each of the three cities. Most of the national studies, essentially policy studies, have generated little construction, and those relating to housing and urban policy, are largely ignored. Port development and major sanitary projects are being carried out, but these are traditional engineering matters moved by foreign aid. There was an exception in the Ministry of Irrigation and its master water plan. This plan, among other things, ranked all the areas identified for reclamation and assigned them Nile water by descending order of soil quality. The water was totally assigned somewhere halfway through the list.

13. Even with the National Transportation Plan, which excludes areas outside the Nile Valley from consideration, the Ministry of Transportation produced this plan without apparent awareness of the plans of the MOD.

14. A governorate is much like a state in the United States as a unit of government between local units and the national government,

except that there are very few revenue-raising powers, much of the staff serve the central government as well, and the president appoints the governor.

APPENDIX: SELECT MAJOR PLANNING STUDIES IN EGYPT, 1973-85

National Studies

Ministry of Development (MOD), Housing for Low Income and Informal Sectors, 1976.

Joint Housing Team (USAID and the Ministries of Housing and Planning), Immediate Action Proposals For Housing, 1976.

Joint Housing Team, Housing and Community Upgrading for Low Income Egyptians, 1977.

USAID and the MOD, National Urban Policy Study, 1982.

UNDP and Ministry of Transportation, National Transportation Plan, circa 1982.

Ministry of Irrigation, National Water Master Plan, 1984.

Ministry of State for Land Reclamation (MOD), National Master Land Plan, circa 1985.

USAID and the Ministry of Agriculture, Agriculture Mechanization Study, 1983.

USAID and the Ministry of Industry, Mineral, Petroleum, and Groundwater Assessment Program, 1986.

Regional Plans and New Lands Studies

UNDP/OPE and the MOD, Suez Canal Regional Plan, 1976, and the following elaborating studies: Suez Regional Industrial Plan, 1978; Suez Regional Tourism Plan, 1978; Building Trade and Manpower Training Study, 1979; Structure Plan for Suez Governorate, 1980; Integrated Agricultural Development Feasibility Study, 1981; Lake Manzala Study, 1981; Solid Waste Management Study, 1981; Human Resources Development Study, 1981.

Japanese International Cooperation Agency (JICA) and the MOD, High Dam Lake Development Planning Study, Phases I and II, 1980.

Government of France (GOF) and the MOD, Regional Plan for the Red Sea Governorate, 1981.

MOD, Regional Plan for the Coastal Zone of the Western Coast, 1976, and the following elaborating studies: North West Coast Structure Plan, 1978; North West Coast Infrastructure Plan, 1980; and North West Coast Tourism Plan, 1981.

USAID and the MOD, Sinai Development Study, Phase I, 1984.

MOD, Sinai North Coast Tourism and Structure Plan, 1986.

MOD, New Valley Regional Development Plan, 1982.

UNDP and Ministry of Planning, Development Plan for Region VIII, circa 1984.

Suez Canal Cities

UNDP/OPE and the MOD, Port Said Area Master Plan, 1976, and the following elaborating studies: UK Overseas Development Ministry (UKODM) and the MOD, Demonstration Projects, 1976-81; UNDP/OPE and the MOD, Urban Land Reclamation Study, 1979; UKOMD and the MOD, Technical Assistance Program, 1979-84.

UNDP/OPE and the MOD, Ismailia Area Master Plan, 1976, and the following elaborating studies: UKODM and the MOD, Demonstration Projects, 1976-81; UKODM and the MOD, Technical Assistance Program, 1979-84.

UNDP/OPE and the MOD, Suez Area Master Plan, 1976, and the following elaborating studies: UKODM and the MOD, Demonstration Projects, 1976; UNDP/OPE and the MOD, Suez Air Quality Study, 1979; UKODM and the MOD, Suez Technical Assistance Program, 1979-84.

New Cities (Free Standing)

The Ministry of Housing and Reconstruction and its successors, the MOD, Tenth of Ramadan New Industrial City, 1979 (under construction since 1979).

MOD, Sadat City, 1980 (initial construction begun 1979).

MOD, New Ameriyah City, 1978 (initial construction begun 1979).

Existing City Plans

GOF and the MOD, Cairo Long Range Development Scheme: Phase I, Strategy Plan, 1982 (Metropolitan Region) and Phase II, Greater Cairo Master Scheme, 1984.

MOD (General Organization for Physical Planning, GOPP), Cairo Satellite Cities: Fifteenth of May, Sixth of October, El Obour, and El Amal, following the Greater Cairo Regional Plan of 1970, circa 1972 to about 1978.

GOF and the MOD, Greater Cairo Food Wholesaling and Distribution Study, 1984.

IBRD and the MOD (GOPP), Lowest Income Housing Study (Cairo, Assuit and Alexandria), 1977.

USAID, the Ministry of Housing and the MOD, Community Upgrading in Cairo and Helwan, 1980-84.

USAID, and the MOD, Informal Housing Sector Study (Cairo and Beni Suef), 1981.

MOD, Planning of the Entrances to the Greater Cairo Area, 1976.

Government of Denmark (DANIDA) and the MOD (GOPP), Master Plan for Giza Governorate, Phase I, 1983.

Governorate of Alexandria, Master Plan for the City of Alexandria, 1984.

Port Development

USAID and the MOD, Development Policy for the Ports of Egypt, 1978.

USAID and the MOD, Master Plan and Infrastructure Development for Damietta Port, 1979.

USAID and the MOD, Port Said Port Rehabilitation Study and Design, 1978.

USAID and the MOD, Port of Suez and Adabiya Master Plan, 1978.

Major Sanitary Projects

USAID and the MOD, Greater Cairo Waterworks Development Programs 1980.

USAID and the MOD, Alexandria Water Development Program, 1979.

UKODM, the MOD, and (later) USAID, Cairo Waste Water Master Plan, 1978.

USIAD and the MOD, Alexandria Waste Water Facilities Master Plan, 1978.

Arab Fund and the MOD, Helwan Sewerage Master Plan, 1978.

USAID and the MOD, Port Said Water and Waste Water Master Plan, 1976; Ismailia Water and Waste Water Master Plan, 1976; Suez Water and Waste Water Master Plan, 1980.

IDA, the Ministry of Housing, and the MOD, Provincial Water Supply and Management Study, 1980.

USAID and the MOD, Management and Tariff Studies for Water and Sewer Systems (Cairo, Helwan, Alexandria, Port Said, Ismailia and Suez), 1980.

GLOSSARY

GOF	-	Government of France
GOPP	-	Egyptian General Organization for Physical Planning
IBRD	-	International Bank for Reconstruction and Development
IDA	-	International Development Association
JICA	-	Japanese International Cooperation Agency
MOD	-	Egyptian Ministry of Development (also known as the Ministry of Housing and Reconstruction, MHOR, Ministry of Development and New Communities, MODANC, etc.)
UKODM	-	United Kingdom Overseas Development Ministry
UNDP	-	United Nations Development Program
UNDP/OPE	-	UNDP Office of Project Execution
USAID	-	United States Agency for International Development.

14
Consultants and Counterparts in Development Planning Projects

A Panel View

This series of views by experts with extensive international field experience addresses the complexities of consultant-counterpart relationships in technical cooperation projects in developing countries. Opinions are expressed about the project context established by donor agencies and host countries, the selection and assignment of foreign consultants and local counterparts, and their involvement in the process of project formulation and execution. It is important to recognize that disparities in the perceptions of consultants and counterparts about their respective roles are quite common, nevertheless there is consensus on the need to review and resolve some obvious points of conflict.

ESTABLISHING STRONG COUNTERPART RELATIONSHIPS OFTEN IGNORED

Malcolm D. Rivkin, Planning Consultant

Developing country assignments for American and European urban professionals generally fall into one of two broad categories:

1. Provision of technical analyses or studies that support the work of the donor agency, such as an assessment of urban conditions prior to a program evaluation, a pre-feasibility study for a loan, or a project evaluation; or that are commissioned by a donor agency to

171

further the development activities of a host country organization, such as a market analysis for housing or a facility.

2. Provision of support for a host country agency that improves that agency's skills and operations, for example establishing a computerized information system, an area plan, a new department, or a training program.

These categories are applicable to both short- and long-term missions. The first represents the kind of work a professional would perform anywhere and can be done in a relatively straightforward fashion depending on availability of data and host country cooperation. The second merits, indeed demands, counterpart support if it is to be at all successful — measured not by reports, but by building in new ways of thinking, new administrative processes, and improved performance.

There are a number of basic difficulties in achieving success, however. In the zeal of international agencies and consultant groups to demonstrate tangible product results, the necessity for establishing strong counterpart relationships is often neglected. In their desire to obtain outside help and please international donors by initiating projects, many host countries often fail to assure proper counterpart availability. Consequently, many of the benefits of foreign assistance are lost because of insufficient discipline on both sides to ensure and maintain appropriate counterpart relationships.

The following recommendations are suggested to redress this situation:

1. International institutions should make a clear distinction in professional assignments among those with a study orientation, those which are to perform institution-building, and those where combinations of these are desired.

2. International institutions should set clear ground rules about counterpart provisions and should not initiate technical support activities until appropriate counterparts are identified and in place. Release of funding should be tied to counterpart availability.

3. In selecting professionals for institution-building assignments, international agencies should make clear that counterpart orientation

and staff training are significant components of the assignment and should recruit personnel with appropriate abilities.

4.   In evaluation of activities that involve institution-building, the performance of counterparts and other support staff should be assessed.

FAIR OPPORTUNITIES FOR LOCAL COUNTERPARTS IS REQUIRED

Mohammad Danisworo, Indonesian Institute of Architects, Bandung, Indonesia

To most Third World countries, the practice of modern urban and rural planning is a relatively new experience, both in terms of the profession and its acceptance within the socio-cultural and political context.   This situation coupled with lack of experience in dealing with massive urbanization and other development pressures has forced planning authorities to adopt planning policies and programs on a trial and error basis.

For more than two decades international donor agencies have provided money and efforts to "help" Third World countries to answer this challenge through various planning activities and projects.   The terms of reference for these projects, in most cases prepared by the donor countries/agencies, normally call for the involvement of "qualified" planners who usually are expected to come from the developed countries.   The terms are also written in the language of the donor country in a manner that makes international tender obligatory and that limits the opportunity for local resources to assume full responsibility in the project team, thus establishing the foreign experts as the principal consultants for the projects.   This places the local counterparts in a less decisive position in the project organization and in many cases causes them to function only as liaison staff, sleeping partners, or even as errand boys.

In this situation, lack of an established body of professional knowledge of planning in a Third World context which could form the basis for operations, coupled with lack of understanding of specific local issues, have made planning efforts more academic and experimental than policy-oriented in nature.   Socio-cultural dimensions are often misunderstood and sometimes even neglected.   Thus, although the product of such planning may be technically superb, the socio-cultural goals associated with the improvement of the quality of life are often disregarded.

Transfer of know-how is basically an introduction of "new" ways of doing things in the host country. As a result, today western methods, and wasteful standards and technologies, are being repeated in the wrong context. Foreign planners, as Donald Appleyard once put it, "carried their bag of tricks around the world to bring out with them wherever they landed".

Inexperience and complications in organization and procedure have made the objectives of planning secondary to the administrative requirements of the process. For instance, the scope, level, extent, and time dimension of planning, together with the billing rates, often have been negotiated and determined by prefixed and unreliable project time schedules and the availability of budget. As a result, assignment of project personnel generally has become defined in terms of the money available.

In my opinion, the fundamental issues of counterpart roles should be viewed in these contexts. As a learning experience, most planning projects have to a certain extent provided young local planners with some skills and technical know-how in planning techniques and methods, but their opportunity for involvement has often been limited to the portion of the work that did not require previous training. The more complex jobs have been done mainly by the senior "skilled" experts brought into the project from outside. If a more innovative approach to the planning and design process had been provided in the terms of reference, the local resources would have been given fairer consideration and weight.

The following recommendations are suggested to overcome these problems:

1. Fair opportunities should be given to local counterparts to play more decisive roles in technical cooperation projects. In areas of planning in which local capabilities are considered sufficient, leadership in the project team should be assigned to local counterparts and foreign experts should contribute their expertise to help solve particular planning problems. Only by this method can the direction of planning be tailored to local requirements.

2. Recipient countries should be involved in the preparation of the terms of reference from the beginning and they should be written in a language understood by all parties concerned so as to prevent any confusion about its content and meaning. For example, in performing feasibility studies there is no way that foreign consultants can communicate with the local populace without the assistance of national counterparts.

3. International agencies together with the host country government should regulate acceptable billing rates, particularly for local counterparts, so as to allow ample time and meaningful participation of well-qualified local counterparts.

4. Transfer of know-how is a two-way operation. Basic to this technology transfer is the eventuality that local counterparts will be able to execute similar projects by themselves. Therefore, to ensure that such technology transfer is to take place, the process of planning should be reflected in the project organization.

COUNTERPARTS SHOULD BE COLLEAGUES AND DIRECT PARTICIPANTS IN PLANNING

Malcolm D. MacNair, Development Consultant

In current theory, the idea of counterparts for international development consultants serves at least three purposes: (1) to enable the transfer of technology to the counterpart; (2) to provide local support and communication so as to maximize the efficient use of expensive expatriate expertise, and minimize work delays; and (3) to provide the expatriate consultants with an understanding of the context and local conditions in order to produce better plans. The notion that technical assistance differs from institution-building in the need and use of counterparts is a false one. There ought to be counterparts in both instances — if such distinctions truly exist — since both activities should aim at achieving plans and policies.

From many years observation of planning in Egypt, I can say that only the third purpose consistently occurred there, although all studies required counterparts and joint venture relationships. This was because the first and second of these purposes were overlooked as linked and mutually generative objectives defining better plans as implementable plans. Many plans and projects in Egypt were not carried out because they were unrealistic in accounting for existing organization and personnel or in devising ways to build local capabilities to plan and execute plans.

There were pressures to minimize the time expended by expatriate experts and yet achieve the rapid completion of planning studies. Host governments want to move quickly, or at least appear to do so, in order to balance political matters. Donors want to minimize costs and meet the demands for accountability. Consultants want to meet contract requirements at a minimum cost (if a lump sum contract) or minimum diversion from the scope of work of the contract (if a cost plus fixed fee contract). Counterparts want to minimize career risks attached to the use of external support that utilizes methods outside of the normal or that challenges their superiors. These pressures impede the development of implementable plans and the building of administrative structures and abilities to carry out plans.

Counterparts ought to be redefined as colleagues and direct participants in planning, particularly in devising ways to schedule, carry out, and coordinate planned activities. The opportunities for technology transfer in planning — as a purely technical activity — are few. On the other hand, the potential for developmental change in realizing plans are many. The transfer should therefore be of administrative skills besides specialized technical expertise. The use of experts should be for longer periods, using fewer experts at any one time, with the carry over of the use of consultants into plan implementation. Although perhaps slower and appearing more expensive — certainly often more frustrating than a technical study — these proposed changes may be more effective and economically efficient in yielding directed and accelerated change and development, particularly in an evolving administrative setting.

DONORS SHOULD PROMOTE DEVELOPMENT OF LOCAL COUNTERPARTS

Viviann Pettersson Gary, US Agency for International Development, Bangkok, Thailand

My experience is that of an employee of a donor agency, thus I am neither a consultant nor a counterpart. I like to view my role as that of a facilitator; unfortunately, both consultants and counterparts often feel the opposite is true.

However burdensome the policies and regulations of donor agencies are, they have become an accepted part of the project design and implementation process in developing countries, basically because of the bottom line: costs. Without donor assistance many of the urban planning projects in developing countries would not be funded. Those that do secure local monies seldom employ foreign consultants.

A basic problem facing international development planning consultants today is that home-based market rates for consultant services are higher than most developing countries are willing or able to pay. Add the cost of travel and *per diem* and the problem is further exacerbated. Thus even with donor provided foreign exchange, developing countries are becoming more and more reluctant to spend their limited development monies on foreign consultants. Increasingly, host countries, at least in Asia, are negotiating with donors for decreased budget line items for expatriate technical assistance. Thus as donor funding decreases, the market for foreign consultants may also wane.

This will lead to a need for a shift in the relationship between donors, foreign consultants, and local counterparts, each of whom plays a valid role. The role of donors is important by virtue of their funds, their ability to push neglected aspects of urban development agendas, and the comparative analysis and personnel expertise they are able to contribute. Foreign consultants have a role not only because of their specializations but because they have a significant benefit of being one step removed from the considerable sways of political agendas. They can be sounding boards for their counterparts. Moreover, if a program is presented and rejected as politically unpalatable, it can be attributed to the foreigners' lack of understanding. Likewise foreign consultants can be used to reassure the powers-that-be that a program previously presented by host country professionals but ignored by policy makers is worthwhile.

Local professional counterparts have the most important role in this triad. They provide the link to the future not only in project implementation, but more importantly in the country's overall urban development process. There is almost always a cadre of competent local planning professionals. In most developing countries the cadre is very small and stretched thin while in others, especially in Asia, the pool of local experts is substantial. The majority of countries, however, lack competent private consulting firms which meet international standards of performance. In most cases local consultants are in fact public sector employees with private practices.

In the project design process there is a typical relationship between donors, host countries, and consultants. Representatives from a donor agency and the host country agree on a problem to be addressed. Each agrees to provide a certain level of financial and human resources to address the problem — a pledge that may or may not be adequate for the task. Local counterparts and donor agency representatives who may or may not be planning professionals are assigned as project managers and monitors. Thus comes the need for consultants who are specialists within the problem area addressed.

While consultants must be hired under a mutually agreed set of guidelines and regulations, the link is often stronger between the consultant and donor than the consultant and host country. International consultants are familiar with the nuances of the bidding and contracting processes established by donor agencies. Local consultants, if they exist, are usually not as experienced with the contracting processes or the performance standards required by donor assisted projects.

The shift that I believe is needed is for donors to actively promote the development of competent local consulting firms and for foreign consultants to merge their forces with local professionals. This shift, which is becoming noticeable in Asia, has numerous advantages. Most important from the development perspective is the potential to strengthen a country's capacity to address its future urban development issues with or without donor assistance. For foreign consultants it provides a mechanism which can result in cost savings and thereby make their services more attractive to host countries in the future. Finally it provides a means whereby local professionals can get on-the-job training in the business of consulting.

This type of shift is not without obstacles. For instance, the salary standards set by donor agencies are often found to be degrading to local professionals because they are based on local wages rather than a given level of competence. Likewise, while foreign consultants frequently subcontract with local firms, they often do not fully utilize the professional capabilities available. Local professional counterparts are often used just as a liaison to set up meetings and collect data rather than as integral members of an analytical effort.

Regardless of the possible difficulties, it seems the time is ripe in many developing nations to shift the focus of consulting efforts to local professionals. Both donors and foreign consultants can encourage this shift. While foreign consultants may be wary of training their future competition, decreasing levels of donor assistance suggest they may in fact broaden their marketability by merging forces with local counterparts.

PROFESSIONAL IMPERIALISM SHOULD BE AVOIDED

Myer R. Wolfe, University of Washington, Seattle

Several lessons emerge from the various comments expressed by the expert consultants.

1. The transfer of technology and knowledge must flow in both directions during the formulation and execution of the program.

2. Not only is institution building *per se* required, but strong emphasis must be placed on the needed competences to be left after a project is completed. The terms of reference and work programs should be jointly formulated by counterparts and consultants at the outset.

3. Not only should interaction between professionals occur before, during and after a project, but there must also be interaction with other indigenous personnel. Indeed, the latter is essential, as professional work cannot operate in a vacuum.

Above all, preparations and agreement beforehand appear to be the critical factor, given that the host community should best know what is needed and that professional imperialism should be carefully avoided. Inherent in this conclusion are the facts that counterparts have their own constituencies to answer to, and that there are bound to be hidden agendas for those whose tenure is comparatively short compared to others who continue in residence.

Acknowledgement

The views expressed by the panelists should not, of course, be attributed to their respective organizations.

New Roles for Communities, Organizations and Planning Educators

15
Community Building –
When People Take Over

John F.C. Turner

Coordinator, Habitat International Council NGO Habitat Project

This paper is based mainly on the results of the Habitat International Council (HIC) NGO Project designed to enable Non-Governmental Organizations (NGOs) and Community-Based Organizations (CBOs) around the world to share their achievements and methods during the United Nations International Year of Shelter for the Homeless. Collectively, the cases demonstrate the ways and means by which the tasks common to all projects and programs can be carried out by people and local groups: self-organization, generating and obtaining credit, acquiring land and tenancy rights, participatory planning and design, the development and use of local materials and skills, the use of local enterprises and self-help for construction and for the management and maintenance of home and neighborhood improvements. They illustrate a greatly enlarged "tool box" of options for preparing projects in ways that suit the particular local situation or for training. They provide a global showcase for raising awareness of the quality and quantity of home and neighborhood building by NGOs and CBOs, and of how much more could be done with government support.

SOME DOCUMENTED CASE STUDIES

Ukanal-Fe: Communal Development in Oussouye. Senegal

The development of a youth group into an association engaged in community support and the installation of community services.

Assisted by an NGO, Ukanal-Fe has built a nursery school, a kinder-garten and a tourist camp, and has created local employment.

Organizing for Change. *Ganeshnager, Poona, India*

The 800 squatters of Ganeshnagar first had to struggle against self-appointed strongmen before the community could take control. Assisted by a local NGO, a development plan was drawn up and the local authority pressured into installing services and sponsoring complementary activities carried out by different committees of the Ganeshnagar community.

Peru's Largest Pueblo Joven. *Villa el Salvador, Lima, Peru*

An organized invasion of private land led to the allocation of govern-ment-owned desert land to the squatters who, after 15 years, have developed a satellite town of some 200,000 inhabitants.

Redd Bama Urban Development Project. *Kebele 41.* *Addis Ababa, Ethiopia*

In an inner-city Kebele (neighborhood level of self-government), the Norwegian NGO Redd Bama initiated an integrated improvement pro-gram, including home improvements, sanitation, employment-gener-ating local enterprise and social development projects. The successful program has shown how low-income communities, in cooperation with the government of a poor country, can be aided to help them-selves.

Project Pouzzolana Chaux, Tourbe (PPCT). *Ruhengeri, Rwanda*

Research and development of a local alternative to expensive portland cement for housing improvement, carried out by a Belgian NGO. It set up a unit producing a binder from local volcanic ashes (*pozzolana*) and lime, and thus generated local employment.

Cooperativo 20 de Junio. *Villa Chaco Chico.* *Cordoba, Argentina*

From a spontaneous settlement of rural migrants, Villa Chaco Chico has emerged as a consolidated settlement. By means of a simple con-struction system, mutual support and their own construction work-force, the tenants' cooperative has built new homes and installed services.

COMMUNITY BUILDING - WHEN PEOPLE TAKE OVER

Women's Construction Collective (WCC). Kingston, Jamaica

The WCC has made a significant contribution to improving the job prospects of women in Kingston. The Collective was initially part of an NGO but has become increasingly independent. It continues to train women and has become successfully involved in various spheres of building.

From Struggle for Survival to Viable Community. Klong Toey, Bangkok, Thailand

A large-scale slum improvement project supported by Government and several NGOs. After a long struggle, the settlers succeeded in negotiating a land-sharing scheme with the port authority, the owner of the land. An improved settlement is under construction.

Low-Cost Sewer Systems by Low-Income People. Orangi Pilot Project, Karachi, Pakistan

In Orangi a drainage system was developed and installed with the help of the OPP at only one-fifth of the cost to the municipality. The inhabitants organized the work and laid and maintained the system partially on their own. This self-organization caused greater partici-pation in political decision taking and further community projects were taken on.

The Uruguayan Experience of Cooperatives. Montevideo, Uruguay

In the inner-city Complejo Bulevar and in Mesa 1 on the city's periphery, the inhabitants have built new homes and carried out accompanying social projects with the technical and logistic support of the Uruguayan Cooperative Movement.

The Palo Alto Struggle. Mexico DF, Mexico

The Cooperativo Palo Alto evolved on occupied land on the outskirts of Mexico City to play a leading role in Mexico. Their 40-year struggle has resulted in new forms of financing, democratic self-administration and new progressive laws.

Guerrero Housing and Services Cooperative. Mexico DF, Mexico

The Cooperative Guerrero in the center of Mexico City has replaced the cramped and decrepit *vecindades* with new buildings with the support of the administration and an NGO. In this way it has demonstrated that renovation of the inner-city for low-income groups is possible.

Kampung Improvement by the People. Banyu Urip, Surabaya, Indonesia

After the inhabitants had been carrying out improvement work for years in self-help schemes, Kampung Banyu Urip was brought into the city's Kampung Improvement Program. Participation of the inhabitants in the planning and carrying out of the works resulted in greatly improved quality and maintenance.

Tarime Rural Development Project (Tardep). Tanzania

A foreign NGO-assisted program in cooperation with a national NGO and with local authorities complemented national housing policies. The program focussed on motivating and training the village people, as well as informing them on the use of local materials and related activities such as reafforestation.

Women Take the Lead in Low-Cost Sanitation. Baldia, Karachi, Pakistan

Encouraged by a woman organizer from an NGO, Baldia's people built their own new soakpits. This intervention produced major shifts in the traditional male-dominated society. Local women gained confidence through their involvement and now manage and take part in home schools and their own health centers.

The HUZA Program. Lusaka, Zambia

Owing to the involvement of HUZA, a national NGO, and its negotiating role between squatters and official bodies, the inhabitants of many squatter settlements were able to carry out significant improvements. Housing standards have improved, training courses and small businesses were set up, and market-gardening extended.

SUMMARY OF THE EVIDENCE

The experiences reported through the HIC Project, together with other sources used for this paper, confirm the view that centrally administered housing supply policies, those based on standardized programs for categories of people, must give way to revolutionary support policies — those based on institutional changes that increase access to locally scarce resources and which protect the right of responsible self-management by people through their own local community-based organizations (CBOs). The reports also confirm the view that the potentials of nongovernmental organizations (NGOs), as well as those of CBOs, are grossly underestimated and underused. It is also concluded that the right kinds of NGO are essential for starting and carrying out the policy changes that must be made if homelessness and loneliness are to be rapidly reduced and finally eliminated.

Facts speak for themselves: with their limited budgets, governments of low-income countries cannot house the low-income majority who cannot afford commercial rents or prices. In high-income countries, where governments have been able to house most of those excluded by the market, management costs are escalating and communities are more often broken down than built up. Many deprived communities, especially in Third World countries, have shown how much more can be done with limited resources when dwellers control the major decisions and are free to make their own contributions in the design, construction or the management of their own homes and neighborhoods. It is when people have not control over, nor responsibility for, the key decisions in the housing process that built environments so often become barriers to community and to personal fulfillment, as well as being a burden on the economy.

The reports on the projects and programs carried out by people, especially those supported by governments or assisted by NGOs, demonstrate how much more can be done with less. There are several cases in which community based action has achieved five times more than either government of the market can provide at the same financial cost.

The 200,000 people of Villa El Salvador on the outskirts of Lima have achieved a level of development that would cost the average resident household 20 years' income. Even when credit is available to low-income people in an unstable economy, few lenders will advance more than three or four years' income. The vital assistance received, in this case, was the allocation of land by their government, some initial utilities and technical assistance, all well within the government's financial and administrative capacity.

Over 300,000 people of Orangi and Baldia in Karachi have installed their own sanitary systems providing, respectively, 95 and 80 per cent of the costs themselves. Improvements to the same standards provided by government and commerce cost about five times more than the actual costs to the communities — and about five times more than they could afford. In these two neighboring cases, each use a different approach and a different technology. Orangi's success is due to technical and very limited financial assistance from a national NGO; Baldia's successful project, also due to national staff, was supported by both national and international sources.

Maintenance can often have an even greater impact on costs and environmental quality than the initial design and construction. The importance of user participation and self-management for maintenance is highlighted by the case of Banyu Urip, a participatory Kampung improvement program in Surabaya. The improvements are in perfect order and the environment is impeccably clean. The residents wash the pavements, keep the drainage channels clear of rubbish, maintain community facilities such as public latrines and laundry areas, and plant public spaces. These and other facts, such as the number of flowers and the attitude of people, may be difficult to quantify or to obtain if they are recorded. But they are obvious to the visitor and clear indications that the community has been built along with and through the material improvements. By contrast, in those Kampungs whose improvements have been planned and implemented without participation, residents wait on the authorities for maintenance and repairs and few or no additional community facilities are installed. Drains are often filled with rubbish, deterioration is often rapid, and many improvements need major repairs after a few years' use. Many of the centrally managed settlement improvement projects in Indonesia are very poorly maintained, needing major repairs after less than ten years' use.

The case histories explain how and why so much more is done with less when people take over, especially when supported by government and assisted by mediating NGOs. It is a fact that the greatest available resource for new home and neighborhood building are those who need them and for improvements and maintenance, those who already have them. How well those resources are used, or if they are used at all, depends on people's freedom to use their unique knowledge or their own needs and priorities; on the choices and decisions they can make for their own experience, skills, time and money; and on their opportunities to demonstrate their own abilities and commitment. Even when people are materially poor, these resources far outweigh those that the state can provide. The good use of land and living space, of materials, of working time and money, all depend as much or even more on iocal knowledge and personal will than they do on professional expertise, industry and social institutions.

So, when large organizations exclude users and residents by monopolizing the production or management of dwellings and related services, they have to replace those personal and local resources that cannot be obtained through market pressures or police powers — and these ways and means of extracting people's contributions also cost time, effort and money, especially when the contributors are dissatisfied and therefore resistant. In societies where people have been led to demand the right to be housed, as distinct from the right of access to resources in order to house themselves, those who succeed pay too little while those excluded usually have to pay too much. On the other hand, when people's voluntary initiative is supported and they are enabled to do what they are able and willing to do for themselves and their neighbors, their contributions usually exceed expectations, they get high returns on their efforts and investment, and the benefits are widely distributed.

The support of local initiative depends most heavily on central government powers to change decision-making structures and access to resources and, but to a lesser extent, on local authority administration. Most actual support, however, is provided by NGOs who must work within existing power structures, who have no direct powers to change legislation and very limited financial resources. The cases show that NGOs are now in the forefront of policy development, despite these limitations. Although rooted in the European and North American charities of the nineteenth century, NGOs are now spearheading attacks on the roots of poverty rather than soothing symptoms of the underlying malaise. The contemporary work of NGOs and the various roles they play reflect an evolution from authoritarian management of projects for the poor, through participatory but still paternalistic community development; from this to community organizing and mediating between self-managing CBOs and the corporate power with which they have to negotiate in order to build or improve their own homes and neighborhoods. A realistic aim for NGOs is also illustrated by the HIC project: to be in the role of advisor and consultant responsible to CBOs able to negotiate on their own account.

People's freedom to do what they are able and willing to do depends on the nature of the limits to what they and their neighbors may do, as well as on access to the essential resources they lack. The liberation and constructive use of the immense and largely wasted resources which people possess depend on three things that only government can control: (1) on rights to make and carry out decisions on where one lives, on one's choice of shelter and the duration of one's tenure, not the right to be housed how and where others decide; (2) on the rules that guarantee these rights for all by setting limits to what may be done — not regulations that lay down lines that people must follow; and (3) on access to the basic resources of land or living space, to skills and time and to appropriate technologies, not on their supply by public or private monopolies.

As Arcot Ramachandran, the Executive Director of the United Nations Center for Human Settlements, stated to the UN Commission meeting in May 1986: "Our agenda for the next ten years must be to find the necessary capacities to apply these enabling strategies: we can only give a guarantee of failure for any other kind of strategy". This agenda seeks changes in the ways decisions are made and carried out, in legislation and in banking and finance. It is therefore political, not a matter of technological gimmickry or of streamlining failed systems. The well-known failures and the little-known successes reported through the HIC project show what has to be done: everyone can join with their neighbors in demanding their rights of access to resources in order to make their own contributions; those in government can stimulate local initiative and inhibit the counter-productive effects of bureaucracy, monopolistic commerce and centralizing technologies; and the communicators can accelerate the change of understanding on which changes of people's demands, commercial and governments' responses depend.

The evidence from the HIC project shows that, as third parties, NGOs are better placed than state or market organizations to carry out five key tasks: (1) to raise public consciousness of the underused capacity of CBOs and NGOs; (2) to stimulate and promote the growing demand by people for the support of local initiative; (3) to assist local groups and communities to organize and in the development of their own projects and programs; (4) to advise governments on the formulation and implementation of support policies; and (5) to undertake the one role that only third parties can perform: to mediate between people and the state.

FURTHER READING

"Building Community: A Third World Case Book," Bertha Turner, ed., available from HIC NGO Habitat Project, AHAS, PO Box 397, London E8, UK.

16
The Guerrero Housing Cooperative: Mexico City

Priscilla Connolly
Universidad Autonoma Metropolitana, Mexico City

CONTEXT

The case of the Guerrero Cooperative is about the experience of a group of tenants living in central Mexico City. They include factory workers, low-grade government or private employees and those self-employed in petty trade and personal services. Belonging to the lower income brackets, they are clearly not free from economic difficulties, especially in the face of high inflation and unemployment. The immediate problems facing the tenants relate to their housing, called *vecindades*. Overcrowded, badly-lit one- or two-roomed dwellings with ill-functioning water supply and drainage, the washing and sanitary facilities are often shared by up to forty families. During the rainy season, the roofs and walls of the old buildings frequently cave in, causing fatal injuries to their occupants.

Despite these disadvantages, this housing in Guerrero was, and still is, relatively cheap. About one-fifth of the population still pays rents which were frozen in the 1940s (that is they were paying — and some still pay — the present equivalent of less than thirty US cents a month). Even the free-market rents remain fairly low: less than one-tenth of the legal minimum wage. Most people have been living in the same dwelling for many years. In fact, many of the inhabitants of the neighborhood have been born and bred there. They have strong family and occupational ties with the area which offers a rich variety of job opportunities, commercial activities and services.

192

The cheap housing, the long-standing community links, and the locational advantages of the neighborhood are the main reason for the tenants' wanting to remain living there. At most, they were willing to change houses but not to change neighborhood. By the mid-1970s, this objective was being increasingly threatened by the workings of the land market. Real-estate investment, both large- and small-scale, was slowly converting the low-cost residential land uses into more profitable buildings: offices, shops and middle-class housing. Land prices in the Guerrero district and similar neighborhoods soared.

From the tenants' point of view, the landlords appeared to be the main enemy. For many years the owners of the *vecindades* had not undertaken any maintenance or repairs, under the pretext, justifiably or not, of the low rents they were receiving. In order to capitalize the potential value of their properties they needed to be rid of the tenants. Various tactics were used for this. Rents suddenly were increased exorbitantly or the houses were simply allowed to fall down. During the Presidential administration of 1970-76, the city authorities actually assisted the landlords in getting rid of their tenants. Removal vans were sent, free of charge, to help move evicted tenants out of the *vecindades* and into new public housing projects built on the City's edge. When this happened, the tenants seldom were told where they were being moved to or how much the new home was going to cost. Many of them stayed in the projects until the first mortgage payment was due, then they would leave either to drift back to the central area or to cheaper housing on the periphery. After 1976, tenants who were evicted either by landlords or because of public building and road-improvement schemes were not even offered the option of a new house on the periphery.

Another important aspect for understanding the Guerrero Cooperative case is the framework for housing finance in Mexico, from 1975 to 1982. On paper, about 70 per cent of the Guerrero neighborhood's resident population was eligible for one or another of the public housing subsidized finance programs. This does not mean that this population had real access to such funding, nor were there sufficient funds for all those who were eligible. The fact is that there was no precedent for public housing finance channelled directly to organized user-groups, still less had such funding ever been used for urban renewal in favor of the original inhabitants.

Since 1982, the effects of the economic crisis in Mexico brought substantial changes to this situation. The real-estate market slumped severely and land speculation was no longer attractive. At the same time, the relation between building costs, interest rate and incomes altered, thus limiting the purchasing power of vast sectors of the population, including residents of the Guerrero neighborhood.

THE TENANTS' ORGANIZATION AND ITS OBJECTIVES

The problems outlined above prompted some tenants in the "Los Angeles" quarter of the Guerrero neighborhood to organize and form a housing cooperative. Under the slogan "We want to die in Guerrero, but not under a pile of rubble", the organization's general aim was to achieve better housing locally without moving to another neighborhood.

The cooperative's more specific objectives were outlined in a "Local Government Plan" which was duly presented to the authorities in August 1976. The Plan, which was aimed at the integrated development of the area, included the following basic programs:

1. Employment and Popular Economy

   - the promotion of economic activities such as cooperative workshops and other kinds of employment;

   - the setting up of consumer cooperatives and subsidized government stores.

2. Land-Use Control

   - counteracting land-speculation by demanding that the neighborhood be declared an "improvement area", with the appropriate land-use and zoning regulations;

   - gradual decontrol compensated by low-cost housing programs.

3. Environment and Community Development

   - emergency cleaning-up program of public spaces and empty lots;

- remodelling of the public square;

- provision of pedestrian streets, multi-use car parks, and a social sports center.

4. Four Housing Programs

- Emergency rehabilitation — roof repairs, shoring up walls, plumbing and closing dangerous rooms;

- Definitive rehabilitation — room extension, lighting and ventilation improvement, introduction of services,

- Minimum housing projects: new buildings for people on minimum wages or unsalaried workers;

- "Housing in process" projects — *vecindad* substitution, financed by existing public housing agencies, for population with access to such programs.

Part of the cooperative's specific objectives included socio-economic and technical studies which were necessary for drawing up the Improvement Plan.

TECHNICAL ASSISTANCE AND NONGOVERNMENTAL
ORGANIZATION SUPPORT

This organizational effort did not happen spontaneously. Strong external support, in the form of social, organizational, legal and financial counselling and technical assistance, was necessary at practically every stage in the cooperative's initial development. Basically, three types of organizations contributed to this process.

First, the incipient tenants' organization was set in motion by a team of priests and seminarians, as part of their pastoral work in the local parish. Second, students and teachers from the university faculties of architecture and social work participated actively in gathering and processing information for the "socio-technical study" carried out by the cooperative. Third, a nongovernmental technical assistance organization provided the impetus, stamina and general coordination through the whole process. The professional team working in this organization advised on cooperative formation, housing finance, urban analysis and planning and credit application, as well as drawing up the housing project plans, obtaining permits and supervising construction.

As an indicator of how intense this NGO technical assistance was at certain stages, it is worth mentioning that during 1976 and 1977 at least six professionals were working full-time on the Guerrero project. This does not include the numerous volunteer workers involved, nor the students and teachers who collaborated on the project as part of their university course work. The same professional team continued to provide advice and assistance to the Guerrero cooperative, under the aegis of a new NGO founded in 1979, the Housing and Urban Studies Center, "CENVI ac".

RESULTS

There are two kinds of results arising out of the Guerrero housing cooperative experience. Its direct results are those affecting the participating tenants and their immediate neighborhood. The indirect results refer to those effects on the aims and actions of other neighborhood organizations, professional bodies and the government, which may be wholly or partially attributed to the cooperative's action. In this case, the indirect results of the Guerrero cooperative are unquestionably more important than its immediate achievements.

Direct Results

During the first two years of its existence, the Guerrero cooperative achieved some of its outlined goals. Evictions were successfully resisted by community solidarity; emergency repairs were carried out and, most importantly, the technical and socio-economic studies leading up to a local improvement plan were elaborated and presented to the authorities and at various public events. For the most part, however, the Plan's contents did not proceed further than the paper it was written on. Although the concept of neighborhood-level urban development planning is stipulated in Mexico City's revised planning legislation of 1976, the legal, financial and administrative instruments for carrying them through, especially when they are proposed by the residents themselves, do not exist.

The goal of developing housing projects in the neighborhood for the cooperative's members was achieved with the help of existing public housing finance mechanisms.   Thus the cooperative's main achievement consists of two housing projects, of 60 and 32 units respectively, which were financed mainly by the Mexican Workers' Housing Fund (INFONOVIT).

It should be stressed that the Guerrero cooperative still functions actively.   It not only administers the housing already built and under construction, but also promotes new projects for its membership.

Indirect Results

The cooperative's indirect results arise out of its demonstration effects which have been transmitted in conferences, seminars, cooperative congresses, the press and professional publications, as well as through informal communications channels.

Important groups influenced by these demonstration effects are the tenants' and squatters' organizations and the popular urban movement.   The Guerrero cooperative's example still inspires many similar neighborhood groups.   This was particularly true in the aftermath of the earthquake, which badly hit the Guerrero neighborhood itself, and other similar central areas occupied by the precarious *vecindades*.   Here, the cooperative's housing projects served not only as an example of what could be achieved, but also represented what the battered but increasingly organized earthquake victims might expect and demand.

The Guerrero cooperative has also contributed to a change in official thinking about the way to tackle inner-city problems in Mexico.   In particular, urban renewal and rehabilitation to the benefit of the original inhabitants is seen not only as feasible, but also as a desirable proposition.   Piecemeal small-scale actions with community participation have replaced the massive bulldoze-and-rebuild form of urban renewal which characterized earlier decades.   It is true that international trends, the debt crisis and the real-estate slump have influenced this change.   However, the fact that some of the key professionals who participated in the initial stages of the Guerrero project went on to occupy top government planning posts undoubtedly contributed to the change in attitudes.

It was not only at the level of ideas and general policies that the influence of the Guerrero experience was felt.   Government policies on urban renewal and rehabilitation also changed in the 1980s. Three different areas of government action are significant.

The first is the reorganization in 1983 of the Popular Housing Fund (FONHAPO) of the state-owned Public Works Bank, and the introduction of innovations which were a radical departure from the previous housing finance programs in Mexico. FONHAPO only grants collective credits, mainly to user organizations and cooperatives. Furthermore, funding for independent technical assistance organizations is included in the credit package, in order to provide the kind of support described in the Guerrero case. On this basis FONHAPO has been financing *vecindad* substitution projects both in Mexico City and in other urban areas, and has been largely instrumental in formulating and implementing the post-earthquake popular housing reconstruction programs.

A second example of the influence of the Guerrero cooperative experience on government policy is the Federal Housing Law passed in 1983. This law states that programs benefiting housing cooperatives and other user associations will be given priority and that technical assistance to such organizations is necessary and should be financed. Notably, the legislation bestowed official recognition on housing cooperatives.

Finally, since the Guerrero cooperative's efforts to funnel public finance into small-scale, user-controlled, urban infill housing projects, there have been many examples of *vecindad* substitution and urban renewal public housing projects implemented on similar lines. The spectacular culmination of this kind of policy is clearly the post-earthquake reconstruction program.

The reconstruction policy for the heavily damaged inner-city *vecindades* hinges around the expropriation of 3569 properties in central Mexico City, mainly comprised of earthquake damaged and/or obsolescent tenements. In the expropriation bill the occupants of these tenements were specified as its beneficiaries. Tenants renting houses in these properties were then automatically granted a right to credit for acquiring a new or rehabilitated house. As the expropriated properties are scattered all over the inner city area, this meant that the program was administered at the level of each plot. Negotiations with the beneficiaries, setting up of representative committees, design of housing projects and establishment of condominium property all take place at this level. Roughly half the program is financed by the World Bank, while the other half comes from the federal treasury. The credit terms are sufficiently modest so that practically all the ex-tenants can meet them. In special hardship cases a reduction of payment is negotiated and additional funding from philanthropic sources is applied. Furthermore, under the Reconstruction Program another of the main difficulties which had previously thwarted the Guerrero cooperative's efforts was eliminated. Existing building and planning regulations, such as density limits and the requirement of at least one parking space per dwelling, were replaced by a much more flexible set of rules.

SOME QUESTIONS FOR DISCUSSION ARISING OUT OF THE GUERRERO CASE STUDY

Looking at the Guerrero experience from the vantage point of more than a decade, four major points for discussion may be identified.

The Demonstration Effects of Small-Scale Pilot Projects

From the analysis of the Guerrero cooperative case, it is clear that limited experiences can contribute to transforming the ways in which housing is provided in a given country. Though the direct effects of the cooperative's effort are fairly minimal, and there is little prospect in the near future of their multiplying significantly, the indirect repercussions of this experience may be felt at a national level.

Any evaluation of these effects must, however, take into account certain special circumstances surrounding the case. For example, one of the leading figures acting on behalf of the first NGO subsequently occupied influential positions in the planning ministry responsible for outlining the national housing policy, and became director of FONHAPO. Also, the upsurge and development of the Popular Urban Movement, coordinated at a national level, has been an important contributing factor in placing local participation in urban policies on the political agenda.

Selecting "Target Population" and Neighborhood Organizations

The help and support afforded to the Guerrero cooperative, in the form of technical assistance and volunteer work, were in no way directed to the poorest and neediest of the population, not even with the city. It is highly unlikely that the results that were obtained would have been possible working with people who had less economic, human and material resources than the participants in this case study. Especially, if such people were not eligible for existing credit packages, it would have proved very difficult to work within the national institutional framework for housing finance, let alone to try and bring about changes in this framework.

On the other hand, two facts about Mexico should be taken into account. First, the housing problem in this country does not only affect the "poorest of the poor", but most of the working classes. If these sectors are not provided with better housing it is difficult to see how decent shelter can be provided for people whose more essential needs such as food are still unresolved. In the second place, if the criteria adopted for supporting housing programs is that they should benefit only the most "needy", then all resources would have to be channelled away from the large cities and into the rural areas where the most extreme poverty is to be found.

The Role of NGO Independent Technical Advice

In terms of economic and human resources, especially those provided by the NGO and other support groups such as the parish and the university, it is evident that the Guerrero cooperative turned out to be very costly. This cost should be measured up against the benefits and, especially, the possibility of reproducing this kind of intensive assistance in the future.

Apart from the benefits derived from the Guerrero case itself it is worth pointing out that the large number of people involved in the project enhances its educational value. Students, bureaucrats, professionals, tenants, priests and, in particular, the NGO team, all learned something from the process, and many later found themselves in a position to usefully apply their acquired knowledge.

The replication of such intensive aid and assistance is difficult to achieve in other situations. Two alternatives might be considered. The first refers to a trend that is already happening among the more enlightened and competent local organizations in which the outside technical advisors are partially replaced by voluntary or paid workers drawn from the ranks of the user group or beneficiaries themselves. This brings the need of training into the forefront, a task that the NGOs could well fulfill, thus multiplying their advisory capacity. The second possibility, which is also being implemented to

a certain extent in Mexico by FONHAPO and other government
agencies, is that the financing body can provide at least some of the
necessary advisory services: informing the potential beneficiaries
about the programs and credit packages, their respective requirements, and how to fill in the forms.   This probably requires some
kind of initial training programs for which the NGO's experience could
be advantageously applied.   Either way, it is clear that independent
NGOs still have an important role to play, even though this role must
change and adapt to evolving circumstances.

Necessary Flexibility in Building and Planning Regulations

One prime obstacle facing the Guerrero cooperative, and indeed
almost all grass-roots user organizations, in their efforts to develop
their own housing solutions, were the rigid and often conflicting
requirements for obtaining the numerous and necessary permits and
licenses.   All housing projects need to be approved and meet the
regulations of various government offices.   It is extremely difficult to
meet all these requirements, for both technical and economic reasons.
Some norms, such as the car-parking requirements, are often impossible to fulfill, given certain cost limits and land prices.   Also the red-
tape involved in acquiring all these permits is exceedingly weari-
some, especially for the unpractised.   On the other hand, the abolition
of all regulatory controls would clearly lead to abuses and lowering of
building and urban standards.   The simplification of regulatory
requirements and bureaucratic procedures, without entailing the
deterioration of the built environment, is therefore a vital condition
of the development of constructive community participation in housing projects.

17
The Emerging Role of Nongovernmental Organizations in Shelter and Urban Development

Richard May, Jr
International Division, American Planning Association

One of the most important trends in international development assistance is the increasing role of nongovernmental organizations (NGOs) in shelter provision and the improvement of housing and urban environments. With growing public commitment and financial contributions the NGOs are forging new relationships with aid agencies, governments and the building industry on behalf of local populations seeking an enhanced participatory role in the planning and development process. Indeed, many NGOs, through the Habitat International Coalition (HIC), the NGO alliance for human settlements, are having a broader impact on shelter and urban development policy issues. This paper considers some of the problems and opportunities facing NGOs.

WHO ARE THE NGOs?

Officially, NGOs are private humanitarian associations which have gained nongovernmental organization status with the major international intergovernmental organizations, such as the United Nations, UNESCO, the Council of Europe, the European Community, and so on. They are obliged to be autonomous, not-for-profit voluntary associations who manage their own resources without State funding as their main source of income. More recently, the term has been loosely applied to international, national and local groups, in both the industrialized and developing countries, who share these characteristics but may not necessarily be accredited by intergovernmental organizations or governments.

Europe and North America can claim the largest number of NGOs; several thousand of them undertake disaster relief, charitable assistance, social work, education, health and activities supporting the social and economic development of the world's poorest people, most of whom live in the southern half of the globe. Some also offer assistance to the underprivileged in developed countries.

Large well-known NGOs include international voluntary agencies such as CARE, OXFAM, and Save the Children, and philanthropic and church-related groups such as the American Friends Service Committee and the Catholic Relief Services. Their activities in developing countries are carried out through field offices, voluntary teams and direct linking relationships with indigenous affiliates.

NGOs indigenous to the Third World, are by contrast, a relatively new phenomenon emerging in increasing numbers. It is estimated that local NGOs may involve as many as 60 million people in Asia, 25 million in Latin America and 12 million in Africa (Borghese, 1987). They constitute a "barefoot revolution" by the millions who have organized themselves into small self-help communities. Many are offshoots of established religious and charitable groups; others have been formed to assist or defy the implementation of government policy. The vast majority, however, are community-based organizations (CBOs), grass roots bodies, housing associations, tenants' and squatters organizations, village committees, and the like formed spontaneously in response to pressing problems and needs. In many countries, universities and technical institutes play a crucial role in the promotion and support of NGOs and CBOs through their research work, consultancy functions and access to communications media.

Most prominent officially recognized NGOs are engaged in development relief activities, food distribution and rural development projects. Major sources of funding include bilateral and multilateral agencies such as the World Bank, the Organization for Cooperation and Development (OECD), the United Nations Development Program and other specialized UN agencies, as well as many national government organizations and private aid agencies. Such funding from private and official government sources has increased considerably over the past two decades. According to a recent report, total grants by the NGOs in OECD nations to developing countries for relief activi-

ties doubled in the 12 years from 1973 to 1985 from approximately
$1.5 billion to almost $3 billion, and about half this amount was
contributed to the NGOs by OECD governments (OECD, 1975, 1985).

International donor support specifically for improvement of
shelter and living conditions for the poor has low priority however
and accounts for only a small proportion of aid flowing to developing
countries. According to one observer, "Few agencies commit more
than 5 per cent of their total budget for urbanization projects and
other forms of public support for improving low-income shelter"
(Blair, 1983). Urban lending by the World Bank, the world's largest
source of multilateral aid for planning, administration and construction in Third World settlements, was only 4.1 per cent in 1981
(World Bank, 1983), and is currently estimated at 6 per cent of total
Bank financing — with no future increases expected in this ratio.
Recent statements in a report to the UN Commission for Human
Settlements indicate that: "In 1982 less than 5 per cent of
concessional aid (including grants) and some 6.5 per cent of official
non-concessional aid was allocated to housing, urban and community
development, water supply, solid waste disposal and the production
of building materials. In total, the annual average for such aid for
the period 1980-84 was some $3 billion." Furthermore, the report
notes the limited impact of funding programs by saying: "Over the
past 20 years, it is unlikely that more than 5 per cent of the urban
populations in developing countries and a considerably lower
proportion of their rural populations have taken part in housing
construction or upgrading project supported by official bilateral or
multilateral agencies" (UN Commission for Human Settlements,

THE HABITAT INTERNATIONAL COALITION (HIC)

One of the leading NGOs advocating the expanded funding and
support of nongovernmental organizations in shelter and urban
development is the Habitat International Coalition (HIC), formerly
known as the Habitat International Council. HIC was founded as a
voluntary organization of professional bodies, voluntary agencies, and
research, scientific and educational institutions committed to the
improvement of human settlements. Its origins were in the Habitat
Forum "peoples' meetings" which ran parallel to the official meeting
of Governments organized by the United Nations Conference on
Human Settlements in Vancouver, Canada, 1976. HIC has worked
closely in recent years with the UN Center for Human Settlements
(Habitat) and its governing body, the UN Commission on Human

Settlements. It has highlighted the plight of the world's poor and sought an increased role for their creative energies in shelter improvement and national development efforts (Searle, 1980).

EMERGING NGO RELATIONSHIPS WITH GOVERNMENT AND THE PRIVATE SECTOR

In the past, serious mistakes were made in shelter assistance programs. Human needs and capacities were often ignored in project planning, and many governments in developing countries over-extended themselves with ambitious programs or loans which placed a heavy debt burden on already strained balance of payments. Difficulties in shelter provision were compounded by high rates of urbanization and population growth, which made urban living conditions much worse. Given these circumstances, interest in the potential role of NGOs as agents of development has grown substantially over the past decade. Their perceived qualities, by comparison with conventional official development assistance, is formidable. They have demonstrated ability to deliver emergency relief and assistance at low cost to many people. They can initiate innovative and flexible responses to emerging social, financial and technical assistance needs at the grass roots level. They have experience implementing small-scale development projects, especially those requiring high levels of involvement and familiarity with local target groups (Dordelman, 1986). These perceived qualities have led governments and international agencies to direct more aid funds through NGOs. They now constitute the third major channel, after bilateral and multilateral aid agencies, through which development assistance is provided to the Third World.

Funding for NGOs and their direct involvement in shelter and urban project planning and execution is only beginning in the field of shelter and urban development, however. This is coming about as there is greater recognition of the abilities of NGOs to work with the poor, to operate effectively as intermediaries between governments and local groups, and to organize and manage participatory projects. NGOs have organized credit unions for the self-employed, small farmers and entrepreneurs, whom the macro-development projects couldn't handle. Often NGOs work in response to development projects that have displaced people or exacerbated problems due to lack of proper analysis of the situation. It is also widely recognized that housing programs at the local level undertaken by community-based organizations with the assistance of NGOs can make a major contribution to national economic growth. Governments are also beginning to realize that the scarcity of funds for shelter provision makes local participation and self-help essential, and that official national shelter agencies cannot by themselves deal with the crisis in shelter provision.

These factors favor the delegation of some responsibilities for shelter and urban improvement to NGOs, especially in situations where local governments are limited in their fiscal and planning capabilities to deliver public services and central governments are viewed as remote and insensitive to community needs. NGOs are well placed to work with communities and can harness their self-help housing skills, attract resources, and upgrade local management capabilities. Increasingly, there is pressure upon national governments by NGOs, and some bilateral and multilateral aid agencies, to delegate responsibilities and project administration to NGOs as a condition of an offer of a loan or grant.

In the current climate of economic thinking, the importance of NGOs can only increase as development policy makers seek to make aid more effective and place greater emphasis on privatization and coordination of aid, and promote policy reform and structural adjustments through policy dialogues, institution building and development management (Drabek, 1986). Officials concerned with cutting costs and making fuller use of the private sector and local entrepreneurial activity in developing countries see NGOs as an important intermediary in such endeavors and recognize their involvement with grass roots participation and local self-reliance as areas of strength (Newman, 1985).

Successful NGO activities in housing have also attracted the attention of governments to their potential as cooperators with the private enterprise sector in the delivery of shelter and infrastructure. NGO projects meet segments of the housing market which many contractors and developers cannot reach. They stimulate the building industry through purchase of materials, vehicles and specialized equipment, and generate employment for skilled and semi-skilled artisans and laborers, upon which much of self-help housebuilding in fact depends. In addition, the improving record of domestic savings in the informal sector and loan repayments by shelter project participants may prove attractive to private-sector finance and banking institutions seeking to invest in low-income housing construction and improvement.

Potential conflict issues do exist however, particularly in regard to land acquisition. Where private land ownership is permitted without restriction, the cost of land becomes inflated to the point of jeopardizing project feasibility. Furthermore, in the urban real-estate market the original low-income beneficiaries of housing frequently face pressures to sell their dwellings to speculators and people with greater means, thus thwarting the aims of the low-income shelter program. Safeguards are required so that solely profit-oriented organizations do not exploit development opportunities to the detriment of equity policies. Experimentation is also necessary with a range of alternative cooperative and not-for-profit solutions to critical housing problems.

NGO ACTIVITIES AND CONSTRAINTS

It is now widely recognized that NGO activities can make important cost-effective contributions to shelter provision and urban improvement. NGOs have been lead agencies in projects encompassing a wide range of activities. These include the acquisition of land for new settlements; raising funds to finance projects from pooled resources or lending agencies; self-help shelter and infrastructure construction; provision and operation of community services; training and employment generation; building material research and manufacture; and advising governments on land tenure legislation and reform of building and development standards.

These efforts have not always been successful, however. Some NGO projects have made the same mistakes as government and international agencies, for example, proceeding with preconceived ideas rather than responding to peoples' needs and priorities. Many NGOs operate in isolation, avoiding cooperation with governmental agencies, and fail to conform with official standards and requirements. The result, too often, is that people are evicted and their homes demolished with great loss of investment, resources, time and effort. NGOs also encounter difficulties in convincing local people that a shelter program is not a charitable gift and requires their own participation in work inputs and honoring loan repayment obligations. Some NGOs give insufficient attention to employment generation, not fully understanding market needs for informal sector activities. Perhaps the area of least success has been in influencing legislation and convincing governments to be more responsive to the needs and capacities of the poor (UNCHS, 1988).

The future of NGOs in developing countries will largely depend on how they help reshape the political mainstream so that their concerns reach a broader constituency. NGOs therefore need to engage in activities which will enable them to obtain and consolidate the political support necessary to realize their goals. The growing shift in emphasis from dependency to self-reliance is giving NGOs new status and governments are beginning to realize that policies of self-reliance cannot be achieved without the participation of local groups.

Nevertheless, there are important constraints to implementing participatory policies and NGO involvement in shelter and community improvement. One inherent problem is the gap between the power structure and the people, extending back to colonial times. Small minorities with wealth and power frequently ignore the poor who often suffer government restrictions rather than receive assistance. When governments have overcome this attitude and are concerned with social problems, their weak economies and need to focus investment on "productive" sectors greatly limits available funding for "unproductive" shelter programs. Furthermore, although NGOs provide an inviting new source of funds, whether by generating their own resources or attracting external funding, the popular participation and community organization elements of their programs are seen by some government officials as posing a potential threat to their power.

Another constraint to an expanding role for NGOs in shelter provision is their apparent *ad hoc* approach to projects and lack of a planned policy framework, as noted in a recent publication of the UN Center for Human Settlements (Habitat): "In nearly all cases the approach of international NGOs is a project-by-project one, rather than related to a particular policy *vis-a-vis* a particular country. By and large these NGOs will begin their negotiations with the government of the developing country in question once the project has been selected, rather than begin negotiations beforehand, with the aim of selecting a project together with the government" UNCHS, 1988).

A further constraint relates to the levels of funding required for NGO activities. According to the UN Center for Human Settlements:

"An increased role of international NGOs will require a growing level of funding. Present trends within the donor community would seem to indicate that there is a desire to channel an increased level of funds through the NGO community and therefore more funds can be expected. However, given the fact that there is not a fast growth in the total level of donor funding, whatever is given extra to the international NGOs will have to come from somewhere: bilateral funding, or the contribution of governments to the international agencies which is becoming less attractive to donor governments for a variety of reasons. There will therefore only be a shift in funding channels and what developing countries receive extra from international NGOs will probably be what they receive less through bilateral programs and the work of international agencies.

"However, if developing countries forge enabling policy partnerships with international NGOs, more funds should become available for the work of such partnerships as governments decrease the utilization of funds for their own direct intervention and instead make resources available for their enabling partners" (UNCHS, 1988).

There is also the question of the limited capacities of international NGOs and their national developing country affiliates to organize local community groups on a large scale. While the experience of NGOs in shelter is impressive, it consists largely of small demonstration projects. The aim in expanding the NGO role is not merely to shift from government and international agency administered projects, but rather to improve shelter and community conditions for the many poor not being reached by present programs. Claims to assign a major share of this responsibility to NGOs implies levels of activity, organization, manpower and capacity far beyond that presently available. There are really very few national NGOs focused on shelter improvement and those that exist are only beginning to develop political strength and experience in project execution. The record of project replication without continued international support is not impressive. When this begins to occur spontaneously as a result of national organizations' response to local requests, and with a need for only minimum outside funding, there will be greater hope for success.

Warning that NGOs will have to focus more on the building up of
indigenous staff capacities and assigning more management responsibility to their national affiliates, an OECD article observes:

"... responsibility is not always delegated by the supporting
agency and training of local people in management and
accounting is often neglected.   The external agency too frequently wants to be seen to "achieve" and hence substitutes for
local efforts.

"Even the most cursory look at funding procedures and forms
for requesting aid for a voluntary project shows that many
agencies focus exclusively on the aid portion of what should be a
combined effort, paying little attention to local inputs.

"Many NGOs of the South are inhibited by a lack of resources, not
so much for projects as for their own institutional development
and overheads..." (Borghese, 1987).

Finally, questions have been raised about the ability of local
partners to cope with an increasing number of projects:

"... the possibility of multiplying international NGO interventions
depends on the capacity of local partners.   Can these local partners take on an increased level of funding and the growing
number of projects that they would be expected to administer as
a result?" (UNCHS, 1988).

STEPS NEEDED

It is now more generally accepted by those involved in the shelter
and development field that the only way to achieve the aim of
replicating activities on a scale even approaching the vast levels of
need is by mobilizing the energies and savings of millions of self-
builders.   Increased responsibility for NGOs and CBOs is a keystone of
this strategy and stress is given to the need for "enabling policies"
and coalitions or partnerships, as outlined in the UN Center's Global
Strategy for Shelter to the Year 2000.   In supporting this view an
influential NGO publication states: "Enabling policies imply that community-based and neighborhood groups serve as planning and
implementing organizations while governments make available — at
reasonable cost — land, infrastructure and those services which self-

help communities are unable to provide themselves" (HIC NGO News, 1988). Governments are urged to implement these enabling policies by entering into coalitions with nongovernmental and community-based organizations.

The Habitat International Coalition (HIC) has played a leading role in convincing UNCHS (Habitat), governments and aid agencies of the necessity for these elements in the global shelter strategy. It has also pressed for acceptance of a set of guiding principles which will limit NGO recognition to only those organizations whose main purpose is to meet the needs of the poor and homeless, and which seek to empower local communities with more control and responsibility for the planning and management of local resources and their distribution. HIC calls for the recognition of the right to housing as a universal human right, embodying protection from eviction, and the recognition by governments of the right of NGOs and CBOs to exist and to organize. Furthermore, it is recommended that each country should have a comprehensive network of active, well-informed NGOs and CBOs established with government consent. Governments are urged to adopt strategies and programs for the strengthening of local government as important agents for the execution of enabling strategies at the local level. Finally, HIC calls upon governments to establish mechanisms at local, regional and national levels whereby NGOs and CBOs can be involved and consulted on shelter and urban development strategies including environmental issues and economic factors influencing shelter policies.

Achieving these goals will require considerable efforts by the NGO community. HIC's successful participation in the activities of the International Year of Shelter for the Homeless (IYSH) suggests that continued and increased involvement of NGOs with governments and aid agencies in shelter improvement and urban development is a real possibility. However, it is clear that drastic changes will have to be made. NGOs will have to raise their level of skills and organizational capacities to undertake integrated projects within a planned framework. They will also have to learn to respond to the need for harmonizing the wide variety of small development programs with the larger-scale state projects thereby rendering them potentially more successful.

As far as HIC itself is concerned, its efforts and credibility have been handicapped until recently by its limited resources and membership, and lack of Third World representation on its Board.

HIC's active membership was mainly academic researchers and lobbying organizations rather than those actively sponsoring and administering international assistance programs in developing countries. Changes in the HIC constitution in 1987 achieved much stronger representation of human settlements NGOs from developing countries. Now, two-thirds of the Board's membership is from Latin America, Asia and Africa, its executive secretary's office is in Mexico City, and among the growing number of its more than 200 NGO affiliates there is an increasing proportion from developing countries. Nevertheless, HIC must broaden its own base, particularly by attracting the active participation of the large national and international NGOs who so far seem to be observing and agreeing, but not enlarging their shelter programs. The reasons for this must first be determined — perhaps they are concerned about the constraints of government and private sector cooperation or the cost of shelter projects. Or perhaps they need advice and staff training by experienced shelter NGOs. Until and unless they enlist their efforts in shelter programs, the global strategy cannot succeed.

Concomitant with the goal of major NGO involvement is the need to expand the level of NGO contributions. There will be no great gain solely from shifting a portion of bilateral and multilateral funding to joint NGO projects. New funds are needed and they must come from increased contributions to NGOs. The large organizations are obviously those most successful in fund raising and therefore a strong effort must be made to attract their interest and participation. Limited support is coming from the leading official multinational and bilateral agencies active in shelter, such as Canada, Denmark, Norway, Sweden, the Netherlands and the USA. For example: "USAID has established special funds in individual countries, administered by local aid missions, which are accessible to local as well as American and international NGOs... The US Congress in 1984 set up an African Development Foundation to support self-help development initiatives, with a budget of $4.5 million for that year. Canada in 1986 earmarked $20 million for use by African NGOs over the next five years" (Borghese, 1987).

Furthermore, there must be a greater focus of energies on encouraging the formation of national and indigenous NGOs in developing countries, and the building up of existing ones for action in the fields of shelter and urban development. European and North American NGOs can play an important role in achieving this goal in collaboration with their partners in developing countries. Support might come from major grant making philanthropic agencies, such as the Ford Foundation, seeking to contribute to local institution building through indigenous NGOs because of a desire to work with disenfranchised groups (Berkman, 1986). Notably, new initiatives for information sharing and networking arrangements are emerging in various parts of the world. A campaign for development activity at the local level in both the industrialized and Third World nations is proposed by the recently formed European Consortium for NGO and Local Authority Joint Action for North-South Cooperation, an affiliation of European-based national and international organizations. Funding for its information, education, linking and project support activities is provided on a 50/50 basis by NGOs/local authorities and the European Community. In the US over one hundred private voluntary organizations (PVOs) have created their own association called "Interaction"; and the International Council of Voluntary Agencies (ICVA) has affiliates from most OECD countries (Borghese, 1987).

CONCLUSION

Nongovernmental organizations are at a critical juncture in shelter and urban development. They can become catalysts for shelter and development activities, innovators in terms of social services and technology transfer, links between indigenous groups and governments, and networks of information and skilled people. By their assistance to CBOs, and by acting as intermediaries between them and national governments, these organizations have tremendous potential to help reverse the trends that have led to the worldwide crisis in shelter. Ultimately, the role of development assistance is to increase people's control over their own lives. This implies having a knowledge base in practice and theory, and a level of stability, both in economic and ecological terms. This is why there is such a strong emphasis on institution-building among development professionals today.

The importance of institution-building, which includes teaching management and leadership skills, and development of strategies for project implementation that assures broad participation by the people affected, is clear. Helping local people to better implement and manage their own projects is almost universally accepted as a crucial direction for development aid to be effective.

Although the potential to make great strides is present, there are certain hurdles that must be surmounted. Among these are the dangers of paternalism, of limited national impact due to over-focussing on the local level, and of far too limited partnerships between governments and local organizations. With appropriate changes, NGOs have an opportunity to have a greater impact by bringing more people into the development process.

International development assistance is a vast field with many players who have different justifications for being involved. NGOs are just one facet of development assistance. However, as their impact and importance grows, the pressure to achieve success increases. NGOs can serve a crucial role in the provision of shelter by operating and succeeding in the gaps where, for one reason or another, governments are incapable of carrying out the kind of work needed.

REFERENCES

Berkman, Herman and Kohn, Christopher, 1986, The role of non-governmental organizations in the development process, Graduate School of Public Administration, New York University.

Blair, Thomas, L., 1983, "Viewpoint. World Bank Lending: End of an Era?," *Cities, International Quarterly on Urban Policy,* Vol. 1, No. 2, November 1983, Butterworth Scientific, Guildford, Surrey.

Borghese, Elena, 1987, Third world development, the role of non-governmental organizations, *OECD Observer*, No. 145, April/May 1987.

Dordelman, Dorothy, 1986, "Toward Long-Term Institutional Development in Africa: The Role of US Non-Governmental Organizations," Cambridge, MA.

Drabek, Anne Gordon, ed., 1986, "The Role of NGOs in the Changing Development Assistance Process," World Development, Washington DC. Special Issue Project Proposal, February 1986. See also Anne Gordon Drabek, ed., 1987, Development alternatives: The challenge for NGOs, supplement to *World Development,* Vol. 15, autumn 1987, Pergamon Press, New York and Oxford.

Habitat International Coalition (HIC) 1988, "NGO News on Human Settlements," No. 1., The Hague, The Netherlands.

Newman, Valerie, 1985, "Diversity in Development," Interaction American Council for Voluntary International Action, New York.

Organization for Economic Co-operation and Development (OECD), 1975, "Development Co-operation: Efforts and Policies of Members of the Development Assistance Committee," Paris.

Organization for Economic Co-operation and Development (OECD), 1985, Twenty-five years of development cooperation: A review, Paris.

Searle, Graham and Hughes, Richard, 1980, "The Habitat Handbook", Earth Resources Research Ltd, London.

The World Bank, 1983, "Learning by Doing: World Bank Lending for Urban Development, 1972-82," The World Bank, Washington, DC.

United Nations Centre for Human Settlements (Habitat), (UNCHS) 1988, "Shelter for the Homeless - The Role of Non-Governmental Organizations" (NGOs), Nairobi, Kenya.

United Nations Commission on Human Settlements, 1987, "Shelter and Services for the Poor - A Call for Action," Report of the Executive Director, HS/C/10/3, 4 February 1987.

18

The Informal Sector and Shelter Provision in the Caribbean

Robert Dubinsky

Project Officer, US Agency for International Development

INTRODUCTION

The Caribbean, like the rest of the developing world, is experiencing dramatic population growth and urbanization. Cities are growing rapidly as population increase continues at a high level and large numbers of people, mostly poor, continue to move to cities in search of employment and a better life. At the same time, Caribbean countries are finding it difficult to create enough jobs to deal with the problems of unemployment, and public debt both internal and external is increasing at a rapid pace.

These trends are taxing the ability of Caribbean governments to deal with the consequences of growth. This is especially reflected in the housing sector where the need for shelter far outpaces the ability of the public or private sector to provide it and where the costs of conventionally built housing is far beyond the reach of most people. The need to expand the supply of public and privately built housing cannot be translated into effective demand because unemployment rates are high and the incomes of most families are low, particularly in relation to the cost of housing. This dysfunction has forced the majority of those seeking housing to secure it outside of the formal housing production system.

Today in most Caribbean islands a majority of the new housing that is constructed is produced outside of the formal building arrangements in which mortgage lenders, government agencies, real estate agents, building contractors, lawyers and architects function. Instead, most households secure their homes through the so called "informal sector", a not uncommon practice even in upper income housing construction.

Perhaps the best way to understand what is meant by the informal sector is first to characterize the formal building arrangements that exist in the Caribbean. Generally, it is a complex process. Developed land is costly. Small, well-located lots, particularly, are rare and expensive. Builders are relatively small and, because of their low volume of production, tend to expect large profits. The costs of securing title to a piece of land, transferring a property, hiring an architect, and paying the fees associated with securing a construction loan or long-term mortgage are high. Building and subdivision standards, inherited from the colonial past, reflect European standards of living and practices and are costly to meet. Many housing finance institutions have limited capital and conservative underwriting criteria. Lenders' traditional clientele are middle- and upper-income borrowers and their practices and requirements make it difficult for lower-income families to qualify for assistance. Generally, the formal housing system does not respond to the needs of lower-income families and effectively excludes them from access to new housing.

By contrast, the "informal sector" operates outside the formal government regulatory system and outside the formal housing production system. In the informal sector people take charge of building their own homes. These homes are modest and may lack indoor plumbing, kitchens or utility connections. In many cases they are built on land the house owner does not own, using self-help techniques. Many people may pay for their houses with cash or borrowings from friends or relatives. Table 1 compares the characteristics of the formal and informal sector building processes.

The purpose of this paper is to review the informal housing sector in the Caribbean, particularly from the perspective of AID's housing mandate, and to suggest some of the implications of informal sector activity for governments and donor agencies. First, the paper describes AID's role in housing and reports on the attention being given to the informal sector. Second, the Caribbean environment is characterized. Next, it presents preliminary findings about the informal sector from AID-sponsored surveys. Finally, it suggests some initiatives that donors and governments might undertake to

Table 1. Characteristics of the Building Process in the Informal and Formal Sectors

	Formal Sector	Informal Sector
Government approvals	Homeowner or builder secures Town Planning and other approvals for the construction of the house	Homeowner does not secure approvals
Land	Homeowner has fee ownership of land and legal evidence of ownership; in many cases land has been surveyed	Homeowner may own land but in high percentage of cases owner rents land, lives on land at no cost or squats on land
Construction materials	Homeowner or contractor purchases new materials from suppliers	Typically, homeowner may buy, steal or collect second-hand materials. In other cases homeowner purchases materials over an extended period of time
Construction method	Homeowner hires contractor to build a house or purchases house and land from builder	Typically, house is partially built by home-owner with assistance from relatives and friends or small contractor
Construction time	Generally 4-6 months	Basic house complete in 6 months to 1 year; finishes and additional rooms completed as funds become available

Site preparation	Homeowner hires contractor or builder who prepares the site	Homeowner prepares site with assistance from relatives and friends
House characteristics	Typically 2-3 bedroom complete house with utility hookups. Hookups are legal	Typically, basic unit with 1-2 bedrooms, with selected utility hookups. In many cases hookups are illegal
Financing	Homeowner secures construction financing from a bank and long-term mortgage from a trust company or building society. High fees are associated with securing mortgage financing	Homeowner finances typically with cash or through use of short-term loans. In many cases funds borrowed from friends, relatives or through informal cooperative-type lending arrangement

maximize the positive features of informal processes and minimize some of its negative aspects.

ROLE OF USAID IN HOUSING

USAID is actively involved in addressing shelter and urbanization problems through its Office of Housing and Urban Programs which operates through seven Regional Housing and Urban Development Offices (RHUDOs) located throughout the world, including one located in Kingston, Jamaica to serve the Caribbean region. AID's principal capital resource for housing is the Housing Guaranty Program, through which about $150 million of assistance is made available annually. AID assistance must be directed to families below the median income in each country.

The Caribbean RHUDO currently has projects underway or under consideration in Jamaica, Haiti, Barbados and St. Lucia. It is financing a broad range of housing and urban development projects and activities that are designed to respond to the diverse needs and problems of the Caribbean. These projects and activities are focussed around a shelter strategy that:

1.   Promotes private sector involvement in building and finances a broad range of shelter solutions and works to focus the role of public agencies on policy, infrastructure and stimulating the private sector.

2.   Encourages policy dialogue between the public and private sectors to develop cooperation and understanding of their roles in shelter and urban development.

3.   Helps public agencies to assess their shelter needs and priorities and to allocate resources accordingly. In this way both the private and public sectors enrich shelter and urban development.

4.   Encourages urban economic development investments by the formal and informal private sector to increase revenue generation and job creation.

AID-assisted Caribbean projects include: financing sites and services schemes and assisting in the development and implementation of the National Shelter Strategy in Jamaica; expanding the availability of financing for home improvements; new houses and land purchases in Barbados; developing a newly established Housing Bank in Haiti; and conducting a comprehensive program of training and technical assistance throughout the region.

RECOGNITION OF THE ROLE AND IMPACT OF THE INFORMAL SECTOR

While many of AID-supported housing efforts have resulted in improved housing in the region, AID has come to realize that a substantial percentage of all housing activity occurs outside the formal institutional and legal system in which AID programs typically function. While the informal system for building houses has always been widespread, it seems to represent an increasing fraction of all the housing being built. Reasons for this include an increasing demand for housing triggered by population growth and rural to urban migration; the rapid escalation in the costs of new housing; the increasing gap between what most people can afford to pay for housing and the cost of housing produced by the formal sector. The impact of these factors is compounded by the inability of governments to develop shelter strategies that can facilitate the development of housing for most people. The target group for AID assistance — those with incomes below the median — more often than not secure their housing through the informal sector.

Given the informal sector's growing importance, the Caribbean RHUDO has sought over the past year to develop a better understanding about how it works and what its implications are for AID-supported programs. AID is now engaged in developing approaches both to facilitate informal sector processes and to eliminate the barriers and constraints that are encouraging people to resort to informal sector practices.

For a number of reasons, Caribbean governments' housing policies and strategies have not given adequate attention to the informal sector. While governments have rarely been able to stop informal development, they have looked upon such shelter as temporary and of poor quality and have judged it to be inefficiently produced and a blighting influence. In recent years, however, governments have begun to change their perspective. Governments have begun to recognize that they lack the resources to provide housing to replace what is being constructed informally or even to build more than a fraction of the housing that is being built informally. History has shown that informally produced housing is improved over time and, in many instances, ends up being of similar quality to that produced by the formal sector.

THE CARIBBEAN ENVIRONMENT

In studying housing development processes in the Caribbean it is important to keep in mind the special demographic, economic and political features of the region which shape housing processes and affect the ability of governments to deal with their housing problems.

Caribbean countries are small and thus their markets for new housing are small. At the same time there is significant diversity in size ranging from countries such as Dominica or St Kitts/Nevis with populations below 100,000 to Barbados with 250,000 people to Dominican Republic with a population size of over six million (see Table 2).

Rapid urbanization is occurring in most Caribbean islands. As Table 2 shows, Haiti, Jamaica and the Dominican Republic have experienced dramatic increases in the proportion of their populations that live in urban places.

There are dramatic differences in the economic circumstances of Caribbean countries. Table 2 shows that Haiti is a very poor country by world standards with a per capita income in 1986 of $342. By contrast Barbados enjoys a per capita income of $3530. Such differences in income are reflected in the housing conditions and circumstances that exist among the islands. Table 2 also shows that overall these Caribbean countries in recent years have suffered from a declining or stagnant per capita income which in turn suggests a declining standard of living.

Among the countries in the region there is wide diversity in the capabilities of their governments and maturation of their housing institutions. At the same time, these governments have limited resources that they can direct to housing and infrastructure development; government programs meet only a small portion of the housing need.

Housing construction costs are very high especially in relation to income levels. This is in part due to the fact that in most Caribbean countries house building has a high import content. For this reason

Table 2.   Per Capita Income, Population and Urbanization Trends of
Selected Caribbean Islands 1960-86

Country	1960	1970	1980	1986
1. Per Capita Income (1986 US$)				
Barbados	1811	3169	3591	3530
Dominican Republic	717	989	1432	1319
Haiti	315	289	387	342
Jamaica	1631	2492	1999	1869
Trinidad/Tobago	1901	2356	3439	2484
2. Population (in thousands)				
Barbados	230	240	249	253
Dominican Republic	3441	4059	5527	6560
Haiti	3575	4231	5016	5427
Jamaica	1682	1869	2133	2340
Trinidad/Tobago	842	955	1094	1181
3. Percentage Urban Population				
Barbados	29	27	—	32
Dominican Republic	33	37	—	53
Haiti	11	20	—	27
Jamaica	23	42	—	48
Trinidad/Tobago	36	53	—	49

Source: Inter-American Development Bank.   Economic and Social Progress in Latin
America (1987 Report), Washington, DC: IADB: 421, 422 and 426.

governments become concerned about stimulating the level of housing construction and investment.   The high cost of housing can also be
attributed to production inefficiencies, high building and infrastructure standards and the small scale of the markets.

High levels of unemployment exist in the Caribbean and for
many wage earners incomes are sporadic or seasonal.   As a result,
many households can only afford minimal outlays for housing.

These circumstances suggest that there is a need for large numbers of new housing units in the Caribbean at a time when public resources are declining, standards of living are stagnant or declining, government resources are severely constrained and housing costs are relatively very high.

AID'S INFORMAL SECTOR RESEARCH PROGRAM

In order to learn more about the housing circumstances of low-income families and understand how the informal sector operates, AID's Caribbean RHUDO is sponsoring a research program on five Caribbean islands — Jamaica (the Kingston metropolitan area), Dominica, St Vincent, Barbados and Haiti. The surveys in the first three islands have been completed and the others are underway. In Kingston a cross-section of low-income neighborhoods was surveyed and in St Vincent and Dominica the study looked at selected infill and urban fringe developments. The studies confirm many notions about the informal sector and provide many useful insights into informal development processes and the housing needs of those who typically make use of the informal sector, low-income families.

As is generally known, our studies found that a large portion of low-income families rent. In fact, the Kingston survey found that 58 per cent of the respondents are renters. This paper does not, however, deal with the implications of this finding but instead focusses on what was learned about the informal sector. The major findings from our surveys and conclusions that can be made about the informal sector are as follows.

The informal sector does dominate new shelter provision while statistically reliable data are not available, the surveys suggest that the informal sector may account for 50-70 per cent of all new housing. Given the dominance of this process in providing new shelter, neither AID nor Caribbean governments can achieve their objectives to improve and augment the housing inventory if they ignore informal activity.

Security of tenure is a relative concept and people are willing to build houses on land they do not legally own. Large numbers of respondents built their homes on land that they may not own or on land where the legal status us unclear. The St Vincent/Dominica studies found 42 per cent of the respondents were squatters. In

Kingston about one-third of the people who owned their dwelling either rented or squatted on the land. This finding suggests that many people may be satisfied with something less than freehold title to a building site.

Many people planned and built their homes over a period of a year or two, building as their financial resources allowed, and were very resourceful in marshalling building resources. In Kingston 75 per cent of respondents who built their homes did so "little little" over an extended period of time. For example, 41 per cent of the Dominica/St Vincent respondents had stockpiled materials for six months or more. Two-thirds of those who built homes, initially built homes with two or less rooms. In all the islands studied very active second-hand building materials markets were discovered. This information indicates that more flexible and open-ended financing arrangements could be helpful to people building on an incremental basis.

Self-help construction methods or a reliance on assistance from friends enabled people to afford to build a new home. Respondents produced housing at a lower cost than could have been provided by the formal sector. In the Dominica/St Vincent study, family members worked on 87 per cent of the units and friends or neighbors (a practice called *coud main*) on 84 per cent. In 78 per cent of the cases the houses were built for less than $4000, about one-third the price of the cheapest formal sector unit. The Kingston study concluded that informal sector builders are building at 20-25 per cent of the cost of the formal sector. This finding suggests that facilitating and improving self-help and cooperative building practices can play an important role in bringing home ownership within the reach of many families.

Use of informal lending arrangements or the financing of house building with cash is widespread. Many respondents built with cash or borrowed from relatives, friends, their employer, and so on. In Dominica/St Vincent 40 per cent of those who borrowed to buy land and 27 per cent of those who borrowed to build their houses got the funds from such informal sources. Loans typically had no fixed repayment period nor was interest charged. Half of the land purchases and two-thirds of the houses that were built were paid for with cash. Only 13 per cent of the Kingston respondents who built their houses stated that they borrowed funds to do so. Given the reluctance of many people to go into debt, improving the availability

of mortgage loans may not be as important as many governments
have believed.

Respondents typically avoided or ignored securing government
approvals or using the legal system in building because of the time
involved or because the requirements for securing the approvals are
inordinately costly.   In St Vincent and Dominica land and buildings
are bought and sold without the use of lawyers.   Less than two-thirds
of the land owners had legal titles.   In Kingston, of those that built
homes, about half did not get permission from government to build.
The implication of these data is that by not addressing the issues of
unrealistic regulatory standards and high professional fees, governments encourage the negative results of extra-legal activities.

In spite of the energy and efforts exhibited by those using
informal processes, the completed studies found that improved
housing is not the highest priority of most of the respondents.   The
Kingston study found that only 20 per cent of the respondents would
spend a $550 windfall on housing.   Sixty-three per cent would spend
it on solidifying or expanding income-generating activities of some
kind.   Given the high levels of unemployment reported and the
marginal nature of many respondents economic circumstances this is
not surprising, but this finding suggests that housing officials need to
give more attention to income generating strategies.

IMPLICATIONS AND POSSIBLE PROGRAMATIC INITIATIVES

We expect that the other studies underway will add additional
insights and depth to our understanding of the informal sector.   It is
clear, however, that the informal sector represents a dynamic, innovative and, in many instances, efficient approach to housing development that is consistent with and responsive to the economic
circumstances of many families, particularly those of low income.   I t
reflects the response of poorly housed households who are unable or
unwilling to operate within the formal sector or who could not afford
formal sector housing solutions.   It results in the production of a large
volume of housing with a minimum of government involvement or
public cost.

On the other hand, because it largely avoids traditional regulatory and approval processes, informal development can produce
housing that is unsafe, environmentally unsound, unhealthy or

unsanitary and can lead to inefficient and wasteful use of land and infrastructure.

Our task is to use the findings of these studies to design more effective programs and strategies to enable the predominant users of the informal processes, low-income families, to improve their housing circumstances. The most promising directions that we are examining are:

1.   Improving peoples' access to affordable building sites and infrastructure. This could be achieved by reducing building and sub-division standards to levels more appropriate for the Caribbean; promoting publicly and privately sponsored sites and services projects; and improving low-income areas' access to water and other services.

2.   Encouraging the development of facilitating institutions that can provide information and assistance to people who build their own shelter. These institutions such as churches, cooperatives, credit unions and other community-based organizations could help such individuals directly or indirectly by interfacing with formal institutions on their behalf. They could help people to build better structures and secure financing, building approvals and utility services. This could be accomplished by making technical assistance and funds available to such organizations.

3.   Promoting reform of building regulatory and land titling systems. Such reforms could improve the overall quality of housing construction, encourage and improve the value of investments in housing, and provide improved security of tenure. To accomplish these reforms governments will have to agree to modify existing practices and systems.

4.   Improving the allocation of scarce government resources for housing so that public sector funds are used to facilitate private sector and informal sector activities rather than to build limited numbers of high-cost housing units. This could be accomplished by governments reallocating their funds and revising their housing strategies.

5.   Focussing more of AID's housing staff's attention on opportunities for promoting urban economic development initiatives to generate additional jobs and improve the ability of people to afford

afford improved housing. AID resources could be targeted to urban economic development projects and projects which maximize the generation of jobs.

While AID's Caribbean RHUDO office has just begun to design projects incorporating these concepts, it is clear that informal sector arrangements will continue to dominate shelter provision in the Caribbean.

19
Changing Training and Education Needs for Human Settlements Planning in Developing Countries

Marja Hoek-Smit
Professor, Fels Center of Government, University of Pennsylvania

INTRODUCTION

Education and training are indispensable instruments in the national development process. Only when developing countries have the knowledge and capacities to analyze their own problems and to create their own solutions can true and sustained progress take place. Yet, training and education programs which ought to be at the forefront of new thinking in development planning often lag behind and are not responsive to changing macro-economic conditions and policy adjustments. Neither do they address the related changes in manpower needed for the implementation and management of policy reforms. This paper explores the changing context of human settlements planning and management in developing countries and its implications for education and training, and for the role of international development agencies and institutions of higher learning.

THE CHANGING CONTEXT OF HUMAN SETTLEMENTS PLANNING

Over the past decade the human settlements problems in developing countries have been viewed very narrowly as one of providing housing for the poor in urban areas through government programs such as site-and-services and slum and squatter upgrading. Central governments play a dominant role in the process of planning and implementation of national housing programs. They have responsi-

bilities for formulating urbanization strategies, designing urban area
master plans, setting standards, assessing needs and affordability
levels, and the provision of finance. In addition, the input of foreign
experts in this process has been considerable since these programs
are often funded partly with foreign loans or grants.

There is a growing realization, however, that these programs
have severe limitations in stimulating sound human settlement
policies which can truly contribute to overall development. The
single focus on government-sponsored housing has failed to address
the wider issues of land, finance, urban economic development and
its linkages with rural development. Bureaucratic complexities and
lack of managerial capacity affect the ability of the programs to reach
the scale required to cope with urban growth. Because central
government agencies have difficulties in understanding the nature of
the local demand for housing and services and the functioning of the
housing market, the poor for whom the project are intended have
seldom benefited directly and often the results are misallocation and
inefficient use of government subsidies. In most cases, central
government agencies have failed to build the community organizations needed for effective planning, implementation and maintenance of housing and services. Local communities and local
government authorities are rarely included in the planning of
projects.

Centrally planned housing programs are also being reconsidered
in light of changing macro-economic circumstances. Many Third
World governments are in the process of making urgently needed
macro-economic adjustments. Housing is one of the public spending
areas often critically analyzed in terms of its effectiveness in contributing to social and economic objectives. The role to be played by
the government, particularly the central government, in the provision
of housing is being questioned in relation to private sector and
community contributions.

Within this overall context of reflection, planning professionals
are searching for alternative ways to improve provision of housing
and related services for urban households and to meet the urban
growth challenge. The emphasis is shifting towards "enabling strategies" in support of locally determined, self-organized and self-
managed settlement programs, according to the UN Center for Human
Settlements (Global Report, 1987). Despite the novelty of the idea
there seems to be consensus on the need for two essential elements:

(1) decentralization of central government offices and devolvement of decision making to local authorities and local communities, and (2) an increased role for the private, cooperative and non-profit sector in the provision of housing and services.

Introducing these elements will have far-reaching consequences for the planning process. Decentralization will require involvement of a much more varied group of individuals and institutions, and a restructuring of responsibilities. Central government's role will be to provide the policy context and develop the main strategies for housing and urban development. Local authorities and communities will undertake local project planning and management, assisted by local officials from various ministries, private sector developers, finance institutions, small contractors, cooperatives and non-profit organizations. This new multi-organizational and multi-sectoral planning environment will change the role of the planner and will require a much more interactive and integrative planning process for human settlement development.

CONSEQUENCES FOR TRAINING AND EDUCATION NEEDS

The existing implementing institutions and administrative procedures in developing countries are in most cases not well-suited to human settlement tasks in the "enabling" sense mentioned above. In many countries, decentralization and development policies are introduced out of political necessity to increase the participation of the population at large in government. There is often reluctance by ministries to truly devolve specific tasks and responsibilities for fear of the financial and political implications. Even in countries where central governments are actively engaged in the decentralization process, the capacities to make planning decisions at the local government and project levels are largely lacking, as are the skills and manpower to organize communities to foster self-reliance.

It seems clear from recent studies that education and training in planning, managerial and communication skills is required if these alternative approaches to human settlement planning are to be successful. (International Assistance Strategy, 1986; Hoek-Smit, 1987) Furthermore, training activities have to be carefully designed to accommodate the highest priority development tasks at various levels of decision making.

At the national policy making level of public and private sector institutions specific training is needed in such subjects as alternative systems of local authority financing and relationships between central and local government, urban economic development, housing finance systems and performance of land and housing markets.

At the local authority and project level, training is needed in general management, economic and financial analysis and evaluation techniques, physical planning, public and municipal finance, the management of public works including the relationship with private sector firms, community involvement and communication skills. Training at this level should emphasize the role of the planner/manager as broker between different government levels, departments, organizations and sectors. It must be performance-oriented and based on practice. The need for training appears to be greatest at this level.

At the community level, training is needed in ways to build viable and self-reliant community organizations, in decision making and planning procedures, in construction and maintenance techniques and in income-generating activities. Community development is often done by nongovernmental organizations (NGOs) and it is therefore critical to include participation by NGO representatives in these training activities.

Major efforts should be directed at enhancing local training capacities in the various subject areas. This requires the development of flexible training and education structures which can respond to changing needs for short-term training and academic education. It also requires the education of teachers, the development of appropriate teaching methods, the introduction of curricula for both long-term educational programs and short-term training, and the preparation of teaching materials and manuals.

Attention should also be given to strengthening education and training programs providing training related to human settlement development, such as planning, engineering, architecture, regional development, public administration and social work. Evaluations of these programs in Third World training institutions, undertaken by the UN Center for Human Settlements and the World Bank, point to severe shortcomings in relation to the changing context of human settlement planning in developing countries (Paul, 1983; Unpublished reports, 1983; Boyce, 1985). One shortcoming is reliance on long-term training at the expense of short-term training possibilities. Short-term refresher training or on-the-job training which could introduce new ideas or solve existing problems in practice are rarely available. In addition, at all levels, the emphasis is on formal technical education and procedural training rather than on organizational and managerial training or the analysis and development of policies. There is no orientation towards the interactive process among various organizations and communities involved in the planning process. Typically, training programs are not interdisciplinary and concentrate on developing one type of professional or officials from one ministerial department. Moreover, training institutions have little contact with actual practice and often teach outdated planning models and methods. Teaching methods favor classroom lecturing rather than exposure to real planning situations and few teachers have actual work experience in the field of human settlement planning. As a result, training often is not designed to meet the demand for specific skills, but reflects only the available skills of the teaching staff.

CHANGES AND CHALLENGES FACING MULTILATERAL
AND BILATERAL AGENCIES

There is a pressing need today for new directions in educational and training programs in developing countries. Two main types of institutions can play a key role in finding solutions to the mismatch between the need for and availability of specific planning capacities: the multilateral and bilateral aid agencies, and American universities and training institutions.

The role of aid agencies in bringing about a positive change in planning education and training for developing countries is crucial. They are close to the realities of day-to-day planning and know the limitations imposed by the lack of skilled manpower at specific levels, and they are familiar with mechanisms to facilitate the rational adjustment of policies and procedures. Moreover, they not only are in a position to discuss these problems with authorities but they control the availability of financing resources needed. Indeed, the international aid and development agencies have sought various ways to address the need for training in response to the new demands. Numerous training activities have been organized in the past, but few of them have been part of a coordinated country or regional approach and funding is often *ad hoc* (Hoek-Smit, 1985, 1986, 1987).

In recent years, however, more systematic approaches to these problems have begun to emerge. Specialized training units created within the UNCHS Habitat, the World Bank, and USAID have initiated training policies and strategies specifying the objectives of training, identifying target groups and urging a coordinated systematic approach to training. They have budgets separate from the operational divisions of their agencies which allow them to initiate training activities for which there is a demand and to undertake networking and clearing-house functions. Regional training courses on urban management issues have been jointly sponsored by these agencies, and recently the UNCHS/World Bank organized a coordinating meeting on training for human settlements in developing countries attended by major North American and European development agencies.

Operational divisions of these agencies are increasingly challenged to include institutional development of line-implementing agencies in their projects through human resource development, skill training and education. This is necessary because operational divisions rarely have adequate information on local and international training resources unless there is a training specialist as part of the team. Also, the project cycle and budget for human settlement projects are usually too tight to include long-term development of institutions and local training centers. For these reasons, there is a strong reliance on short-term training inputs directed to a limited number of project staff, often to the exclusion of nonproject staff in implementing agencies and local authorities.

Enhancing long-term training capacities at the local level is the area of greatest challenge. What is required is closer collaboration and linking of financial, human and information resources between training units and operational divisions of international agencies. It also requires a different type of consulting arrangement than the prevalent short-term training contracts with private sector consulting firms.

THE ROLE OF US UNIVERSITIES AND TRAINING INSTITUTIONS

Large numbers of middle- and high-level professionals from developing countries are currently training in US educational institutions in PhD and Masters programs related to urban planning. Many of them receive scholarships from USAID and international agencies. However, these scholarships are seldom targeted to the changing needs in human settlement planning. They are usually allocated by departments that are not directly involved with human settlement problems. It is therefore up to the university planning departments to make their programs relevant to foreign students.

Increasingly, many planning departments depend on students from developing countries to compensate for a declining enrollment of US students. As Sanyal (1985) shows, the question of how the needs of foreign students can best be met is an issue of considerable debate. Should a separate curriculum be presented, focussed on the issues and situations in developing countries? Or should a comparative perspective on relevant approaches to human settlement development worldwide be offered both to students from the US and from developing countries?

I believe the special needs of Masters and PhD students from developing countries will soon appear to be essentially similar to those of US students, particularly the need for better interaction between university-level teaching and the actual complex practice of planning and management of urban areas. Many teaching programs lack an orientation towards the communities which are supposed to partake and benefit from planning efforts. These needs are as relevant in curriculum development for US students as they are for students from developing countries. They are only more pronounced and less easy to solve in the case of foreign students. In large part, the lack of a connection with the reality of planning reflects the shortage of professionals at US planning departments who have on-going planning experience and awareness of new developments in developing countries. Tenure regulations at US universities may be a contributing factor since tenure decisions give little credit to part-time practical experience, even in professional programs. Consequently, practitioners who aspire to tenured positions usually must give full-time attention to teaching.

Greater consideration should be given to the distinction between PhD- and Masters-level students in relation to their future positions.

Upon their return PhD recipients are likely to work either at high-level policy oriented positions or as university professors in planning-related departments. Since the importance of PhD-level education in planning for students of developing countries is often underestimated and considered a luxury or waste, it does not always receive appropriate care and scrutiny. To overcome this is important because PhD-level students are crucial to the institution of a sound policy context for human settlement development and to the operation of teaching and training programs in their countries. PhD students who intend to teach upon their return to their country should receive special assistance during their studies in such areas as curriculum development, training, and education skills.

Masters-level students are more likely to be concerned with the planning, organization and management of urban development strategies. Their curriculum should reflect this orientation and include courses in management, municipal and public finance, economic and financial analysis and evaluation techniques, and communications skills. The teaching of planning models and planning theory should emphasize the interactive and multi-organizational nature of the planning process and should be grounded in practice.

Organizing short-term specialized courses for mid-career, high-level professionals is another area where US universities and training institutions have an important contribution to make (Hoek-Smit, 1986). This is often the only means through which high-level officials can avail themselves of the expertise of top experts in relevant subject areas. The fee-structure prevalent in the international consulting world prevents access to this level of experts for general technical assistance, particularly in areas of finance and management where the domestic market for consulting inputs is at its highest. Independently organized training programs can circumvent this problem because they are paid for through tuition. In order to reach a wider audience and to make training more economically viable, short-term courses can be given in selected developing countries and linked with efforts to develop local training capacities in particular subject areas. As is the case with the longer-term educational programs, nearly all trainees in such short-term programs are funded by international development agencies.

An evaluation of short-term courses at US universities shows that their success depends on several factors, including the involvement of top-level experts as trainers, the selection of participants who can influence policy decisions in their own country, and the availability of follow-up inputs by the sponsoring agency which can be facilitated by the enrollment in these courses of staff members of international development agencies. A training program in Housing Finance for Developing Countries recently initiated at the Fels Center of Government, University of Pennsylvania shows the potential of such programs. It uses experts from Fels, the Wharton School, the international development agencies and the private sector. A few other courses along similar lines have been organized at other US institutions of higher learning.

US universities or training institutions can also contribute to the enhancement of education and training for human settlement planning through direct assistance to teaching institutions in developing countries in program and curriculum development, the training of faculty, and the provision of teaching resources and information. Although many linking and exchange programs exist, the exchange programs are mostly set up to provide learning experiences for faculty and students from the US rather than to enhance the training capacities of the developing country. However, when used professionally and selectively, such programs can have a positive impact on the educational capacities of developing countries.

DEVELOPING TRAINING CAPACITIES IN DEVELOPING COUNTRIES

Where do these various inputs of international agencies and US teaching programs lead us? The agencies are very much aware of the need for a change in the existing training context if developing countries are to deal creatively with their human settlements problems. Their achievements in short-term training activities have shown the potential of training as a tool for development. These agencies, however, have two fundamental problems which must be addressed — a project approach to training and a lack of funding for the development of training institutions.

Universities and training institutions in the US, although engaged in training large numbers of planning professionals from developing countries, have difficulties establishing links with practice which could make this education more relevant. They have little supportive institutional contact with training and education institutions in developing countries. The most immediate need in the present assistance patterns is the development of local training capacities and institutions which can deal with the growing demand for mid-level planners and managers.

Increasingly, requests for longer-term technical assistance are made by line-implementing institutions and government training institutions of developing countries for the development of in-service training programs. It is to be expected that educational programs in professional planning and urban management in developing countries will be called upon to deliver more relevant pre-service education. The vast and varied resources of US training institutions could be utilized for this purpose. It will be up to the international development agencies and private foundations to find funding for this type of institution building and to establish creative ways to provide coordinated inputs into this long-term process of development of training institutions, using local and international training resources from both the public and private sectors.

If there is one lesson that we have learned from the experience with training and teaching programs in human settlement planning in both less and more developed countries, it is the need to incorporate from the outset a combination of teaching, research and consulting activities as well as an involvement in policy dialogue (Fischer, 1986). This combination of activities is relevant whether we deal with in-service training programs or with academic and administrative pre-service educational programs. Combining teaching with practice is important for both teacher and trainee. The training itself will be more up-to-date and the trainee, if exposed to practice at various levels, will be able to analyze and understand the various components and ambiguities of the policy making and planning process.

There are many barriers to overcome in achieving this combination of activities in a training institution. The task of doing so requires a flexible attitude by school administrations regarding appointments and teaching schedules, an appreciation of the value of outside research and consulting in promotion opportunities, and a realistic attitude towards remuneration from outside consulting. Impediments with respect to such flexibility are often particularly strong in governmental public administration schools or training programs in developing countries. Yet is is clear that a rigid separation of academic, training and practical planning activities has had a negative impact on planning programs in the US and developing countries alike. We should be particularly careful to avoid exporting to developing countries the shortcomings which hamper our own teaching programs in the planning and management of human settlements.

REFERENCES

Boyce, Charles P., 1985, "Suggested Networks of Human Settlements Training Institutions in Developing Countries," draft report UNCHS.

Fischer, Frederick, E., 1986, "Training is Development, Pass It On," unpublished paper, RHUDO/ESA, Nairobi.

Hoek-Smit, Marja, C., Reports on the Evaluations of USAID/PRE/H Regional Training Programs in Latin America, 1985, East and Southern Africa, 1986, and West and Central Africa, 1987.

Hoek-Smit, Marja, C., 1986, "Documentation and Evaluation of Selected US-Based Training Programs in Shelter and Urban Development," USAID, Washington, DC.

Hoek-Smit, Marja, C., 1987, "IYSH Demonstration Project on Training and Information in Low-Income Shelter Programs: The Case of Sri Lanka's Million Houses Programme," UNCHS, Nairobi.

International Assistance Strategy for Human Settlements Training in Developing Countries, 1986, Draft report of a working group of the steering committee meeting of UNCHS June 3-5, 1986. See also various background papers submitted to the Steering Committee.

Paul, Samuel, 1983, "Training for Public Administration and Management in Developing Countries: A Review," World Bank Staff Working Paper No. 584, Washington, DC.

Sanyal, Bishwapriya, 1985, One world view of education in the US universities for planning in developing countries, Paper for the Association of Collegiate Schools of Planning Annual Conference, November 2, 1985, Atlanta, Georgia.

United Nations Centre for Human Settlements (Habitat), 1987, "Global Report on Human Settlements," Oxford University Press, Oxford.

Unpublished Reports by the Economic Commission for Latin America, the Economic Commission for Africa, and the Economic and Social Commission for Asia and the Pacific on Human Settlements Training Facilities in each region 1983.

20
Mobilizing Development Expertise for Human Settlements Planning

Thomas L. Blair
Associate, The Martin Centre for Architectural and Urban Studies, University of Cambridge, England

The urban planning and development problems of Third World nations in Africa, Asia and Latin America and the Caribbean are massive, urgent, and complex. Now, after a long saga of trial and error with state-centered, sectoral planning and *ad hoc* urban development projects, and in the face of declining public resources and eroding standards of living and high costs of housing, new goals have been established to create a strengthened urban public service with greater public participation. Briefly, these are:

"1. Formulating national urban policies which define and harmonize at all levels of government the inter-related goals of economic and physical land use planning with social equity and better living and working conditions, particularly for the poorest communities.

2. Organizing, strengthening and coordinating public and private institutional, financial, legislative and administrative management systems for plan implementation.

3. Designing integrated affordable and implementable programs and projects which utilize the initiatives, resources and creative energies of local communities, individuals and nongovernmental organizations.

4. Creating broad education and training programs for policy makers, professionals, project personnel and community implementors.

5. Establishing appropriate and cost-effective national and international arrangements for technical cooperation, information transfer, and trade and aid which take into account the needs and priorities of urbanization in developing countries" (Blair, 1984).

Achieving these goals will require new educational philosophies and experimentation with a wide range of programs in universities and technical institutes, on the job, and in communities. How to mobilize an appropriate educational response to these demands is a major challenge to scholars and professionals, and to governments, lending agencies and voluntary organizations.

This paper deals with the requirements for change in Western and US planning education and training programs if they are to meet the emergent needs of Third World planners, overseas consultants and US students preparing for planning careers with an international dimension. Many Third World students are following planning courses in the USA, although these are sometimes of doubtful relevance to their own countries' problems. But many planning academics and professionals have little interest in developing countries or in providing courses appropriate to overseas personnel. These issues must be addressed, and training and education programs must be reoriented to new goals. Fortunately, there is an increasing awareness of the lessons to be learned from urban innovations on a global scale, and there is greater recognition of the importance of the international transfer and diffusion of planning ideas and technology through education systems, aid agencies and consultancy practice. In America a number of notable efforts have taken place which merit a brief comment.

Within the planning profession, the conferences and networking activities of the International Division of the American Planning Association (APA) have established its role as the key professional organization liaising with planners outside the US and in the developing countries. More than 400 members and several thousand subscribers read its newsletter *Interplan,* and information exchange is encouraged through directories of national and international planning institutes and associations, job opportunities for American planners in developing countries, and urban and regional research sources in East and West Europe.

In the academic sphere, the Association of Collegiate Schools of Planning (ACSP) has given a prominent place to international plan-

ning and development in recent meetings. The 1985 annual conference in Atlanta, Georgia, under the presidency of Professor Jay
Chatterjee, examined the status of planning education for developing
nations in US universities. The discussions traced the recent historical background and highlighted some of the requirements for training, research and international cooperation necessary to meet future
needs. Through the efforts of Professor Herman Berkman and
Richard May, Jr, the ACPS and the International Division of APA have
formed a special interest group in international development and
planning which enrols the support of more than 40 schools and 200
individuals. This impetus was sustained in the 1988 ACSP conference
at the State University of Buffalo, co-chaired by Professors Jay Stein
and Ibrahim Jammal, and conference themes covered such topics as
development theories; case studies on planning in developing
countries; plan formulation; technology transfer; the role of foreign
aid and multi-national companies in national and regional development; national urbanization policies; the informal sector in economic
development; and population and migration theories.

These initiatives emerge at a time of growing awareness that
significant changes have taken place in the composition of the
student population in US planning schools. The proportion of international students increased from under 10 per cent in 1970 to over
20 per cent in the early 1980s. Now an estimated 1000 out of 4500
planning students are from abroad, and to a large extent from developing countries. While total enrollment shrank by 13 per cent
between 1977 and 1982, due to a decline in domestic students, there
was an estimated increase in foreign student enrolments to 61 per
cent during the same period. Though this trend may have slackened
since 1982, the share of international students in US planning schools
may be as high as 25-30 per cent, and in some places, like Howard
University School of Architecture and Planning, may be between 50
and 65 per cent.

Rising enrolments of Third World students in US planning
schools suggest a challenge to go beyond *ad hoc* arrangements and
special programs in Third World planning and restructure the whole
planning education and training curricula. Many questions need to
be raised and answered. Why do foreign students come to the USA
and Western countries for planning education? What is the special
role of developing areas courses? How will the presence of foreign
students assist in the scholarly search for general principles of planning theory and practice? Where does restructuring the curriculum

begin, and where does it end? Who teaches the teachers, and trains the trainers? Why should planning schools be in the business of providing planning education for Third World nations? Who gains, and who loses?

PROBLEM AREAS

In embarking on this very necessary re-examination of planning programs there is no room for naivety. Few planning schools have recognized the contribution that Third World-focussed courses can make to their educational curricula. Where such courses exist they are too often fragmented appendages to mainstream faculty interests, confused about their role and functions, and without power in the academic arena. With few exceptions, planning schools are ill-equipped to teach courses in the international urban planning and development field. Typical problems are: a weak information and knowledge base, a shortage of appropriately educated and experienced staff, inadequate support and cooperation from other university departments and administration, and poor links with aid agencies, consultants and nongovernmental organizations. In short, they lack the critical mass of scholars and resources required for credibility in the field. With additional resources, perhaps some of these problems might be alleviated. More serious are the inherent problems of conventional planning academia and curricula which lack an international perspective and are conceptually and ideologically Euro-centric.

It must also be recognized that even where education and training programs, and professional consultancies, are ostensibly dedicated to Third World planning, they are inextricably entwined with, and shaped by, the general flow of urban development capital and finance and the products of planning ideas, procedures and technology around the world. The needs and demands of the major nations and agencies in the planning export and import trade are in many ways the deciding factors in this international transfer, and the role of international agencies and lending institutions, governments and multinational companies is all pervasive (Blair, 1983).

On a much more mundane but no less important level for our purposes, a large segment of planning academics and professionals exhibit very little interest in planning in developing countries, if one is to judge from a report by the book review editors of the *Journal of the American Planning Association.* According to Eugenie Birch and Peter Salins, the subject of planning and policy in developing countries accounts for only 4 per cent of publications received, 2 per cent of the publications reviewed in the journal, and ranks 2 per cent on an index of professional relevance as reflected in the test specifications for the examinations of the American Institute of Certified Planners. The editors report that: "If we examine each distribution's bottom six (subjects) we find agreement only to the extent that no one seems very interested in demographic and spatial analysis or planning in developing countries" (Birch and Salins, 1984).

Nevertheless, there must be many planners who respect and recall the notable efforts of past scholars and practitioners to define the problems and opportunities for a credible, if not yet popular, field of study. The case for a comparative science of urbanization and planning and socially responsive education, training, research and professional practice was put years ago by Charles Abrams in his widely read book *Housing in the Modern World* (Abrams, 1966). John Friedmann identified the shortcomings and the potential contributions of the US planning export trade — its planning programs for students from developing countries and its overseas consultancies (Friedmann, 1969). Alison Ravetz has helped us to understand the dual concern of excolonial nations to send students abroad and to set up their own centers of education, albeit with external advice (Ravetz, 1980). Too often, however, the stated purpose of narrowing the technical, scientific and normative gap between the industrialized countries and the Third World results in new forms of economic and cultural dependence (Mazrui, 1978). Or, put more subtly by Ian Masser: "There is a probability that the experience of the exporting country may be imposed upon the receiving country without adequate reference to their basic needs", and this likelihood is further increased where there are conditions attached to the technical assistance provided (Masser, 1986).

Yet, as Lloyd Rodwin points out, the attraction of foreign students to the USA still persists, even though it is well known that Western planning education programs, with few exceptions "are geared to their own needs — that is, their own economic circumstances, institutional requirements, technology and values". And he concludes: "Small wonder, then, that there is increasing dissatisfaction with things as they are, and an increasing number of corrosive evaluations of these educational offerings" (Rodwin, 1981). This of course brings us back to the concerns of this paper: the need for change in response to new challenges. Rodwin suggests that the theory and practice of city planning and development should be thoroughly reviewed, and new elements such as implementation, management, and the politics of institutional change, among others, should be introduced into planning programs, in both the developed and developing countries. Consideration will also have to be given to creating new education and training programs within traditional schools of architecture planning and engineering, and if necessary, outside them.

WHAT NEEDS TO BE DONE

Understandably, developing countries seek to build up their institutional planning capability to respond to the pressures of urbanization and development, and in the short-term they seek to train currently active planners and educate future planners in in-country programs and abroad. Organizing an appropriate Western educational response calls for re-thinking and re-orienting the design and execution of curricula for both short-course training and professional and graduate programs in planning. What is crucial, however, is that instructional programs be conceived within a human settlements perspective. This concept identifies the improvement of human settlements, cities, towns and villages, as the focal point for overall national economic and social development, and gives prime consideration to seeking harmonious growth and development of settlements, and settlements networks and linkages, to improve the quality of life of those who are or who will live in them. Education and training programs should form part of a broad strategy to increase the supply of manpower capable of translating development plans into successful implementation in an integrated, environmentally sound manner. All international agencies and governments agencies involved in the transfer of resources, skills and technology should promote and institute education and training programs in the field of human settlements and include it as a central component of their policies, programs and projects (Blair, 1985).

The human settlements approach is of special value in short-course training programs for mid-career training and upgrading of skills of urban public officials, planners and technical staff from developing countries. It is they who have the best opportunities to make effective improvements in public policy and action for urban settlements in an era of relentless pressures. The main course themes should be urban management, development operations, and policy integration and implementation. The broad objectives would be to rapidly assist participants to gain an understanding and operational grasp of the interdependence of managerial functions with the improvement of settlement planning in cities. The need for integration of settlement policies with economic, social and physical planning policies at all levels of government would be particularly emphasized, and special consideration given to the formulation of policies and strategies for meeting the needs of the poorest persons and communities.

Personnel in other sectors providing urban services or involved in urban projects can also benefit from this approach. This seems quite clear from my recent experience with a training program for Third World managers of water supply and sanitation enterprises. As governments and international lending agencies seek sector efficiency in water supply and sanitation they will necessarily focus on improving arrangements for planning and implementing urban infrastructure investment programs and address local government manpower development needs, particularly with respect to management, planning, programing, and budgeting. There will be a need to decentralize planning and decision making and to integrate water supply and sanitation with other sectoral urban development projects. Appropriately designed short-course training programs can assist water managers in the identification of the human settlements criteria for decision making with regard particularly to local judgements about priorities across and within sectors. Participants can be encouraged to integrate technical and scientific insights with an understanding of organizational structures, community dynamics and the decision making process.

The application and exploration of human settlements principles and methods will of course vary at different levels of education; nevertheless, graduate programs at Master's, professional and doctoral levels in US planning schools can also benefit from the introduction of a human settlements approach and courses on planning in developing countries. At George Washington University as a visiting professor in 1985, I introduced a Third World-focussed course on international urban planning. "Home students" were in the overwhelming majority, but the course also attracted a number of foreign students. Students were encouraged to examine and compare theories and approaches to urban planning and development in various Third World countries. Case studies were presented dealing with urban and regional development projects, slum and squatter upgrading, and sites and services in cities as diverse as Nairobi, Kingston, New Delhi, Mexico City and Lusaka. Staff and students joined with invited speakers from the World Bank, the US Agency for International Development, overseas consultancies and nongovernmental organizations in open seminars on such topics as urban development assessment methods and urban project implementation in Third World cities: practice and theory; and lessons from experience.

Looking forward to the future it is possible to envisage a full program of courses in human settlements planning at doctoral, professional Masters and undergraduate degree level. The courses would help students to broaden their intellectual conception of the nature, purpose, processes and techniques of urban planning, to consider the relevance of the international experience for the improvement of the effectiveness of US planning, and to gain an appreciation of the skills and requirements for work in urban planning and development abroad. There is also the possibility of establishing university centers of excellence in human settlements planning and development to provide a unique environment where scholars, professionals and policy makers meet to explore, define and clarify issues and problems across disciplinary, cultural and national boundaries.

BUILDING A CRITICAL MASS OF DEVELOPMENT EXPERTISE

Upon arriving at the University of Virginia as a Commonwealth visiting professor in the autumn of 1987, I was struck by the apparent richness and variety of the University's resources, but wondered how much of these were relevant to scholars and researchers with international and Third World interests. I therefore encouraged my students to join me in compiling a directory of resources for international development studies, including library resources, relevant graduate and undergraduate courses, university centers and institutes, and of faculty and staff members with relevant expertise, as well as student organizations and local area groups interested in Third World development (University of Virginia, Resources for International Development, 1988). Information was included on dozens of subjects under broad headings ranging from agriculture, disaster planning, energy and housing, to pests, population and transportation. Courses listed ranged from Anthropological Perspectives on the Third World to Immigration Law to Nursing Strategies for Rural Community Health. Faculty and staff expertise on

Third World development ranged from former diplomats to physicians specializing in infant mortality and tropical disease, to engineers, scientists, business analysts and humanities scholars, many of whom have active field experience in Third World countries.

We discovered that though a scientist studying tropical weather patterns, an economist studying worldwide businesses, and a health librarian cataloging nutrition information all have one thing in common — expertise that could be useful in improving living conditions in the world's poorest countries — the chances are, none could refer you to the others' work. The directory was therefore useful as a collegial information source and a starting point for those interested in expanding the University's contribution to international development studies. It was also a great help to students seeking to increase their understanding and awareness of international development issues.

Many educational institutions with planning schools have resource bases which can be tapped to support the building up of the requisite critical mass of expertise in international planning and development studies. International development studies, as conceived here, is concerned with the integrative analysis of the urban and regional planning, design and development process in Third World nations. It combines emerging principles in architecture, city planning and the built environment professions with cultural and scientific as well as economic, social and public policy aspects. It recognizes that Third World Cities provide the setting for a global laboratory of urban experience and could be a proving-ground for studies of theory, practice and public policy which transcend the discrete boundaries of conventional academic disciplines. Preparing an inventory and assessment of these resources can be a first step towards greater cross-disciplinary cooperation and the launching of a new and potent kind of intellectual sharing of benefit to planning schools and to emerging societies.

CONCLUSIONS

In sum, the major conclusions which emerge from this paper are:

1. Planning education and training in Western and US schools has failed to realize its potential for aiding Third World countries meet the crisis of urbanization and development in two crucial areas of

need — urban public sector planning and management personnel, and graduate-level city planners.

2. The conceptual and pedagogical causes of failure are unlikely to be remedied merely by strenuous efforts to strengthen planning education and training.

3. Instead, planning education and training should be re-oriented towards a more holistic view which argues for a human settlements perspective on planning and urban development, and planning schools should seek to collaborate with other university departments interested in international development studies, in theory and practice.

REFERENCES

Abrams, Charles, 1966, "Housing in the Modern World: Man's Struggle for Shelter in an Urbanizing World," Faber and Faber, London.
Birch, Eugenie, and Salins, Peter, 1984, Book reviews, *Journal of the American Planning Association*, Autumn 1984:526-27.
Blair, Thomas. L., 1983, Viewpoint: World Bank urban lending: End of an era?, *Cities, The International Quarterly on Urban Policy*, Vol. 1, No. 2, November 1983.
Blair, Thomas, L., ed., 1984, "Urban Innovation Abroad: Problem Cities in Search of Solutions," Plenum Press, New York and London.
Blair, Thomas, L., ed., 1985, "Strengthening Urban Management: International Perspectives and Issues," Plenum Press: New York and London: 194-215.
Friedmann, John, 1969, Intention and reality: American planning overseas, *Journal of the American Institute of Planners*, Vol. 35:187-94.
Masser, Ian, 1986, The transferability of planning experience between countries, *in:* Ian Masser and Richard Williams, eds., "Learning From Other Countries," Geo Books, Norwich.
Mazrui, Ali, 1978, "Political Values and the Educated Classes in Africa," Heinemann, London.
Ravetz, Alison, 1980, "Remaking Cities," Croom Helm, London.
Rodwin, Lloyd, 1981, "Cities and City Planning, Chapter 12, Plenum Press, New York and London.
University of Virginia, Resources for International Development, Blair, Thomas L., Chief Compiler 1988. A directory of information sources, relevant curricula, and leading specialists

the University of Virginia and in the Charlottesville area.   Copies are available from the Secretary, Division of Urban and Environmental Planning, School of Architecture, the University of Virginia, Campbell Hall, Charlottesville, VA 22903, USA.

Epilogue:
Exploring the Parameters of
Global Survival

21

The Urbanization Revolution:
Challenge to International Planning

Noel J. Brown
Director, New York Liaison Office
United Nations Environment Program (UNEP)

If we are to move beyond the conditioned reflex of merely groping toward a more coherent system of effective environmental management, then complex long-term forecasting as well as better planning would seem to be an urgent policy imperative. Those of us who are concerned with this question are convinced that the planning community must become more aggressively involved in the long-term environmental issues affecting urban development planning. This is a problem that the world community can ill afford to neglect, and the reasons should be quite clear.

In a matter of months the human population will top the five billion mark. Clearly this is the first time in the history of the world when planet Earth would be obliged to bear a "human cargo" of this magnitude, and the implications are likely to be staggering. Somehow, something new will have been added to human history as the sheer weight of our numbers slowly transforms us into what some have already termed active "geological agents". This, when coupled with the fact that scientific and technological advances are making us increasingly part of the earth system and a force of earth change, can only lead us to wonder whether a new chapter is not about to be opened in the human story, a chapter which will demand some fundamental rethinking of our approaches to production, consumption, wealth and waste, as well as land and space use.

While at this stage much remains in the realm of speculation with a variety of different groups playing with a combination of different numbers, at least one effect would seem likely; namely, that the people-to-resource ratios will be changed negatively, and that severe strains will be placed on various resource systems. The fundamental question seems to be how much longer can the natural order sustain human enterprise, and at what point will our irrepressible drive for progress propel us to transgress those outer limits on which the stability of the biosphere depends? Will we develop the capacity to manage the effects? Can technological or social adjustments work out appropriate solutions?

Two years ago, Lester Brown, President of the Worldwatch Institute, in commenting on the population dimension of what was then called the African crisis, observed that if the pressures of human activities continue to produce the kind of profound environmental changes that we are witnessing in Africa then we are on the edge of an unfolding human drama on a scale that has no precedent, one that the world is not well prepared to manage. Perhaps more than at any other time in our history we are obliged to confront a fundamental new reality that mankind must now become a responsible architect of the kind of world we will inhabit in a future that is likely to be largely man-made. Our design for living will thus have to be refashioned as we invent an environmental strategy for our survival.

Such a strategy will clearly have to be based on a better understanding and acceptance of the parameters of our survival and the establishment of habitations and production systems geared to sustain existence within these parameters. The intellectual task of this endeavor is complicated by the fact that this is indeed a transitional era, a macro-transition if you will. The world is somehow suspended between a declining industrial order and an emerging era in which electronic technology is likely to be the prime determinant of our psychological and social orientation. It is an era in which, as some have suggested, reality seems more fluid than solid — where humanity is buffeted on every side by persistent and gigantic changes the pace and dimensions of which have not only impaired the human faculty of observation and comprehension, but has created a sense of disorientation and dislocation in all of our fundamental relationships — play, work, others, self, nature — and have left humanity in a state that Alvin Toffler correctly calls FUTURE SHOCK — the Disease of Change, the premature arrival of the future for which we are unprepared.

In order to encourage such preparation with the necessary planning in at least one sector the United Nations convened the Conference on Human Settlements at Vancouver in 1976. This, it was felt, could provide an appropriate point of departure since human settlements represent those primary and secondary environments where life may be enhanced or polluted at its source. The conference brought into very sharp focus the fact that within the gambit of human settlements we are best able to observe and assess the effects of the collision between the man-made and the natural environments from which we will face the challenge of a man-made future.

As everyone knows, the human species has shown a remarkable capacity to adapt to a radical series of changes in the natural environment and has over several thousand years of experience behind it. Will we, however, be as successful in adapting to the man-made environment? We don't know. We have no precedents, only symptoms. Perhaps the most vivid expression of our adaptive dilemma is to be found in the cities — as yet the most comprehensive statement of the man-made environment. The cities, once the highest expression of our civilization and a celebration of the human genius, have now begun to raise serious questions about the future of the man-made environment and about our capacity to adapt to them, much less to manage them. At issue is not merely the question of size, or the ability to deliver accustomed services. There are many more subtle issues whose main dimensions are only now becoming apparent and which will increasingly demand our priority attention.

Let me illustrate the point. Some years ago I was fascinated with a report about the problem of trees and urban stress. Scientists in greenhouses and laboratories throughout the United States were working on major experiments to evolve trees able to withstand city conditions. Trees such as sycamores, maples, elms, walnuts, weeping willows and lindens just could not stand up to prevailing urban conditions. The hazards to their survival include: root restriction caused by asphalt and concrete; root contamination caused by salt used to melt snow; root suffocation by leaking gas from underground pipelines; insufficient water supply; pollution; damage from automobiles and bicycles; and over-exposure to high intensity lighting caused by safe-street concerns. Somehow the trees suffered a kind of "urban burnout".

What is significant here is that scientists were trying to create certain strains of flora adaptable to city conditions. But what about people? We know that urbanization, industrialization and expanding technology are rapidly altering the environment and exposing man to massive amounts of harmful pollutants. Yet we know of no laboratories where scientists are working with equal vigor to create a comparable human strain capable of coping with similar urban stress. One example is indoor air pollution, where in a number of instances the air quality indoors in worse than that outdoors — a phenomenon prosaically characterized as the "sick building" syndrome.

This is how UNEP described the situation recently: The safety of the space we call home is being threatened by numerous pollutants, affecting the health, work and well being of people all over the world. In its various forms, indoor air pollution is a problem which leaves few unscathed. The rich and the poor, developing and developed country people, office workers and mothers, are all affected in different ways. Rural women who spend hours cooking over wood-fuelled stoves are sometimes inhaling tar at levels which smoking at least 100 high-tar cigarettes a day would produce. Families in more affluent societies, who insulate their homes against the cold to cut fuel costs, are unwittingly trapping radioactive radon, which may be responsible for as many as 20,000 cancer deaths every year in the Unites States alone.

While these are extreme examples, innumerable other pollutants are found indoors, such as cigarette smoke, formaldehyde used in furniture and other products, asbestos, carbon monoxide from stove and other combustion, household pesticides, as well as household paint strippers and similar other pollutants used indoors, not to mention a variety of bacteria agents. Those most vulnerable are women, children, and the elderly who spend more time indoors. But, on average, everyone is inside 90 per cent of the time, at least in the temperate zone. So while we pay little attention to our indoor environments, they are also crucial to our health and well being.

My interest is not to join the chorus of doomsters but merely to bring more clearly into focus certain specific issues of the man-made environment - of which the city is perhaps the clearest expression — and which looms very large in our future. And this is not an incidental matter. Throughout the world an urban revolution is fully underway and urban populations are growing much faster than the overall demographic increases — in many instances three or four times as fast. The policy implications of these developments are at least statistically clear. Within a generation urban populations will increase by about two billion — 1.5 billion of whom will live in developing countries. Within the same period, the number of cities of more than one million people will increase from fewer than 100 twenty-five years ago to almost 300 five years from now. Sixteen of these will be megacities of more than ten million people, and the developing countries will have at least ten of the sixteen.

Coupled with this is the fact that shanty towns are multiplying three times as fast as socially acceptable suburbs. These towns are without essential services or health, education or employment — towns which deprive the inhabitants of even the essentials of a basic standard of life, so much so that many can never experience the happiness of simply being alive. Moreover, with this proliferation will come perhaps an increase in squalor, and misery, crime and social despair unequalled in the history of mankind. Increases of these orders of magnitude will undoubtedly have vast national and global repercussions and will require major policy decisions and massive investment programs merely to maintain the present unacceptable level of life — let alone improve them. For one thing, such patterns of human concentration as they outstrip the capacity to provide productive employment will also outstrip the basic human resources and services required to maintain the population, resulting in such environmental problems as inadequate and contaminated water supplies, inadequate health services, as well as inadequate sewage and waste recycling and disposal systems.

Perhaps the dilemma of municipal waste disposal was best reflected in a report which the New York Times characterized as "the garbage without a country" wherein a barge had been floating for several days in the Gulf of Mexico with 3100 tons of garbage from a Long Island community. The problem — how to legally dispose of it. North Carolina didn't want it, Louisiana didn't want it, and if it were dumped in the oceans it would be a violation of US law which forbids ocean dumping. When last seen it was headed for Mexican waters where Mexican reaction will be the same, only this time in Spanish: "Gringo garbage — No".

Even if we were able to find a safe haven for the disposal of this cargo, this will not be the end of the story — perhaps just the opening chapters of a new challenge facing developed and developing countries, which are now being tempted to become the dumpsites for municipal waste from the US. During the last two months the UNEP office in New York has been contracted by two Caribbean govern-

ments that have asked for expert advice on how they should respond
to very lucrative offers to become disposal sites for municipal waste
from the US — a price which on the surface would seem like a good
bargain. This is likely to continue as the build-up of waste continues
in many of the developed countries.

Waste should not be allowed to fend for itself, nor in
desperation surreptitiously seek safe haven.   This is a challenge
facing the planning community since, regardless of the current
situation, waste will one day have to become a resource of the future.
In the order of nature there is no such thing as waste.   Every being
has a function and every product has a use.   Perhaps the time has
come, at least in this one instance, to follow Nature's example and try
to find effective ways of managing and disposing of waste.

Under these circumstances there can be little doubt that satis-
factory human settlements have become a central human challenge
and an emerging global priority.   That is why we in the United
Nations Environment Program welcomed the declaration of 1987 as
the International year of Shelter for the Homeless.   Not only did it
provide yet another opportunity to urge governments and people
everywhere to demonstrate a new commitment to the shelter needs
of the poor and disadvantaged and to move with more vigor towards
effective solutions to the problem of homelessness, but it may also
give new impetus for our ability to reassess our land and space use
as well as the uses and abuses to which humanity has exposed planet
Earth.

That is why we in the environmental community consider the
year also to be a "metaphor" — an environmental metaphor for what
humanity is doing to its planetary home, the shelter for all living
things.   The issue is not simply the viability and productivity of the
planet but its very habitability, the livability issue.   This was the
concern that led scientists to launch an International
Geosphere/Biosphere Program, perhaps the largest science project in
history, one which involved the scientific community and institutions
from over 71 countries.   This project is designed to study the Earth
system as a whole — an interconnected whole — and is premised on
the fact that human activity is destined to change the global
environment in the next century more radically than at any other
time since the Ice Age, posing new and fundamental threats to the
future of the human experiment.   Among the more manifest of these
changes are the warming of the climate from increased carbon

dioxide in the atmosphere leading to ozone depletion and the so-called "greenhouse" effect; the intensive clearing of forests for agriculture; the increasing release of chemicals never before encountered in nature into the atmosphere, soils and ocean, as well as the little understood manner in which these changes tend to interact.

I suggest that as a result of these activities, human beings are not simply ecosystems creatures but now are ecosystems shapers, or mishapers. We have now become a force with nature, with a capacity to affect natural systems in a way that is unknown in the entire human experience. And this will require a new set of responsibilities, new procedures and mechanisms for managing our life-support system.

Everywhere the evidence is clear and conclusive. Whether our concern is with the degradation of our atmospheric resources through the problems of global warming, acid rain, and ozone depletion, or the biotic impoverishment, such things as the loss of genetic diversity, desertification and deforestation, or more importantly, genetic engineering and the risk of introduced species; not to mention the problems of industrial safety. They are problems which create challenges of a new order of magnitude to you in the planning community.

For those of us who tend to despair, I would like to reassure you that, after all, this planet was not designed with us in mind, but we have it within our power to make it a safe and agreeable habitat where sustained livability is possible. But we must plan now, and this should be a central challenge to your profession.

Contributors

Thomas L. Blair
Associate, The Martin Centre for
  Architectural and Urban Studies
The University of Cambridge
Cambridge, UK

Marc Boleat
Secretary-General
International Union of Building Societies
  and Savings Associations
London, UK

Noel J. Brown
Director
New York Liaison Office
United Nations Environment Program
New York, NY, USA

Priscilla Connolly
Universidad Autonoma Metropolitana
Azoapotzalco & Centro de la Vivienda y
  Estudios Urbanos
Mexico City, Mexico

John M. Courtney
Senior Urban Planner
Water Supply and Urban Development
  Department
The World Bank
Washington, DC, USA

Mohammad Danisworo
Vice President
Indonesia Institute of Architects
Bandung, Indonesia

Peter L. Doan
Program in Urban and Regional Studies
Cornell University
Ithaca, NY, USA

Robert Dubinsky
Project Officer
US Agency for International Development
Kingston, Jamaica

Salah El-Shakhs
Professor of Urban Planning and Policy
  Development
Rutgers University
New Brunswick, NJ, USA

Hugh A. Evans
Assistant Professor
School of Urban Planning
University of Southern California
Los Angeles, CA, USA

Viviann Petterson Gary
Regional Housing and Urban Development
US Agency for International Development
Bangkok, Thailand

Marja Hoek-Smit
Professor
Fels Center of Government
University of Pennsylvania
Philadelphia, PA, USA

261

Darshan Johal
Assistant to the Executive Director
UN Centre for Human Settlements
  (Habitat)
Nairobi, Kenya

Peter M. Kimm
Director
Office of Housing and Urban Programs
US Agency for International Development
Washington, DC, USA

G. Thomas Kingsley
Principal Research Associate
The Urban Institute
Washington, DC, USA

David B. Lewis
Associate Professor
Department of City and Regional Planning
Cornell University
Ithaca, NY, USA

Malcolm D. MacNair
Development Consultant
Pt Hasfarm Dian Konsultan
Jakarta, Indonesia

Richard May, Jr
Urban and Regional Planner and Chair,
  International Division
American Planning Association
Washington, DC, USA

Bertrand Renaud
Housing Finance Advisor
The World Bank
Washington, DC, USA

Malcolm D. Rivkin
Rivkin Associates
Bethesda, MD, USA

Dennis A. Rondinelli
Senior Policy Analyst
Office of International Programs
Research Triangle Institute
Research Triangle Park, NC, USA

Israel Stollman
Executive Director
American Planning Association
Washington, DC, USA

Raymond J. Struyk
Principal Research Associate
The Urban Institute
Washington, DC, USA

John F. C. Turner
Coordinator
Habitat International Council NGO
  Habitat Project
London, UK

Myer R. Wolfe
Professor and Dean Emeritus
College of Architecture and Urban
  Planning
University of Washington
Seattle, WA, USA

Conference Seminar Organizers

Eric Carlson
Institute for Public Administration
New York, NY, USA

John M. Geraghty
International Affairs
Department of Housing and Urban
    Development
Washington, DC, USA

Ibrahim Jammal
Director
Center for Comparative Studies in
    Development Planning
School of Architecture and
    Environmental Design
State University of Buffalo
Buffalo, NY, USA

Barclay G. Jones
Professor
City and Regional Planning
Cornell University
Ithaca, NY, USA

John P. Keith
President
Regional Plan Association
New York, NY, USA

Linda Lacey
Assistant Professor
City and Regional Planning
University of North Carolina
Chapel Hill, NC, USA

David Mammen
Institute of Public Administration
New York, NY, USA

Howard Sumka
Office of Housing and Urban Programs
US Agency for International Development
Washington, DC, USA

John F. C. Turner
Habitat International Council NGO
    Habitat Project
London, UK

Myer R. Wolfe
Professor and Dean Emeritus
College of Architecture and Urban
    Planning
University of Washington
Seattle, WA, USA

Index